The Vascular Cambium

The Vascular Cambium

Its development and activity

W. R. Philipson
Josephine M. Ward
B. G. Butterfield
University of Canterbury
New Zealand

CHAPMAN & HALL LTD
11 New Fetter Lane · London EC4

First published 1971
© *W. R. Philipson, J. M. Ward,*
B. G. Butterfield 1971
Set in Photon Times 10 on 12 pt. by
Richard Clay (The Chaucer Press), Ltd., Bungay, Suffolk
and printed in Great Britain by
Fletcher & Son, Ltd., Norwich, Norfolk

SBN 412 10400 8

Distributed in the U.S.A.
by Barnes & Noble, Inc.

Contents

Preface

The lateral meristem known as the vascular cambium is the source of the greater bulk of plant material, and the tissues it produces are some of the most complex found in plants. It is from the cambium that wood, one of the world's greatest natural products, is derived. The development, action, and control of the cambium has therefore received a good deal of attention, but the literature is scattered, and no comprehensive account is available.

The purpose of this book is to gather together what is known of the morphology, development, activity, and physiological control of the cambium and to provide for each topic a comprehensive but select bibliography including important original sources. The scope of the book includes the production of cells within the cambium, but their differentiation into secondary vascular tissue is considered only in so far as it relates to events within the cambium.

A striking characteristic of the cambium is its plasticity. It may arise in any position in any tissue except the epidermis, and it may function in a variety of ways. The cambium is a complex meristem which produces tissues in highly specific patterns. That this balanced morphogenesis is largely under hormonal control is only now being demonstrated experimentally. Apart from its great interest to botanists, the cambium is of importance in the practical fields of forestry, and timber production and utilization. Many of the properties of wood depend on the length of its cells, but in spite of extensive work in this field, the relationship of tracheid and fibre length to the activity of the cambium has not been well understood. A consideration of the processes of division and elongation of the cambial initials is essential to a full understanding of these properties of wood.

We are grateful for permission to reproduce illustrations from a number of publications. The sources from which they come have been acknowledged in the captions.

1
The nature of
the cambium

As the growing points of the shoots of seed plants are small in comparison with the diameter of the mature stem, some means of thickening must be present. Considerable increase in diameter occurs during the primary phase of growth. This is effected by cell division and cell enlargement while extension growth and tissue differentiation are still active. In many plants, particularly in most monocotyledons, stem thickening is exclusively by this means.

Stems which thicken during primary growth are typically cylindrical, as in many palms. If they taper, this is due to changes in the degree of primary thickening. On the other hand, in most dicotyledons and gymnosperms stem thickening is continued into the secondary phase of growth. That is to say, though the stems have ceased to elongate, they still increase in thickness. As this is a continuing process, the older the stem is, the thicker it becomes. Consequently, stems with a means of secondary thickening characteristically are tapered from base to extremity. The additional tissues which increase the bulk of the stem are usually vascular, consisting of both xylem and phloem elements. They are derived from a lateral meristem which encircles the stem and which arises between the xylem and the phloem of the primary vascular system. This meristem is known as the vascular cambium, or usually, briefly, as the cambium. Bannan (1955) has outlined the history of the use of the word *cambium*. It was originally applied, by Grew in 1682, to a refined sap which became elaborated as it moved from the bark to the wood. German botanists applied the term to several meristematic areas, which they recognized as cellular, but it was Sanio (1863) who distinguished between the procambial cylinder and the vascular cambium, so that the term cambium came to be restricted as at present to lateral meristems. It is now recognized that the vascular cambium is not uniform, but consists of cambial initials (cells which perpetuate the meristem) and their undifferentiated derivatives. The entire region is termed the cambial zone or simply the cambium.

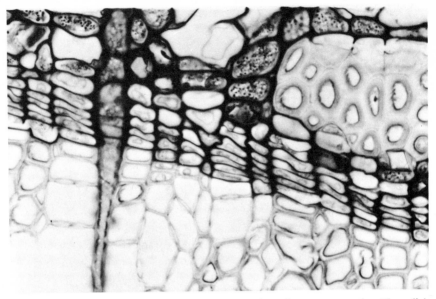

Fig. 1.1 Cambial zone of *Populus* × *euamericana* cv. Robusta in transverse section. The radial files evident in the zone become partly obscured by distortions during differentiation of phloem (above) and xylem (below), ×325. (Photograph by courtesy of Dr J. Stahel, Zürich.)

The cambial zone normally consists of an unbroken cylinder of undifferentiated cells, arranged in radial files (Fig. 1.1). These files extend into the mature secondary vascular tissues, although the pattern may here be obscured by changes in cell dimensions during maturation. Undifferentiated cells to the outside of the cambial initials are known as phloem mother cells, those to the inside as xylem mother cells. These mother cells may be undergoing periclinal divisions. The transition from the cambial zone to differentiating phloem and xylem is gradual, especially in the active cambia of dicotyledons. It may then be difficult to define the radial extent of the cambial zone. Catesson (1964) made a careful attempt to assess the limits of the cambial zone. She considered it best to reserve this term for a truly meristematic region. This she defined as those cell layers in which the ribonucleic acid content is greatest, in which mitoses are most abundant, and which are distinguished in sections by their narrow cells with thin walls. This cambial zone, when active, is separated from the phloem and xylem by a more or less wide band of derivatives in process of differentiation, which, though they may retain the capacity to divide, are noticeably less rich in ribonucleic acid, have thicker walls, and are broader radially.

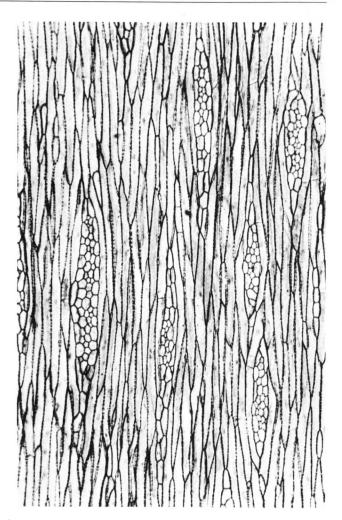

Fig. 1.2 Cambium of *Fuchsia excorticata* (J. R. & G. Forst.) Linn.f. in tangential view, showing differentiation into fusiform and ray initials, × 100.

The cambial initials are usually of two distinct types, fusiform initials and ray initials (Fig. 1.2). Fusiform initials are radially and tangentially short but vertically elongated cells with tapering ends. They give rise to the vertical elements of the secondary xylem and phloem – the tracheids, vessels, sieve tubes, companion cells, fibres, and parenchyma. Fusiform initials are generally larger in all dimensions in gymnosperms than in dicotyledons. Bailey (1920), in a survey of thirteen gymnospermous and fifty-four dicotyledonous species, found a range in length of 0·7–5·0 mm in gymnosperms and

0·14–1·62 mm in dicotyledons. Ray initials are small, almost isodiametric cells which produce the horizontal elements, or rays, of the secondary vascular system.

The relative numbers of fusiform and ray initials and their arrangement in the initial layer varies considerably. Ray initials may occupy over half the cambial circumference or, at the other extreme, they may be very few or absent. They are found interspersed in groups among the fusiform initials. The groups of ray initials vary in width and length in different taxonomic groups and even within the same plant. Where a group is several cells wide the outer cells may differ in shape from the inner ones. The pattern of ray and fusiform initials in the cambium is carried through to the secondary vascular tissues, and may be clearly seen in the mature wood. A classification of types, and an account of salient lines of structural specialization in the wood rays of dicotyledons, has been given by Kribs (1935) (Chapter 3).

Two basic patterns may be distinguished in the arrangement of fusiform initials in the cambium. In gymnosperms and most dicotyledons the ends of vertically adjacent initials overlap in a random manner and the cambium is said to be non-storeyed. Some structurally specialized dicotyledons, on the other hand, have storeyed cambia in which the fusiform initials are arranged in horizontal tiers (Fig. 5.1 and Chapter 5). The cambial initials are usually envisaged as lying in a ring around the stem, each initial being equidistant from the stem centre. However, Newman (1956) found that in *Pinus radiata* D. Don, initials in adjacent files lay at different distances from the stem centre. This is undoubtedly correlated with the fact, also observed by Newman, that adjacent initials do not necessarily add to the xylem or phloem simultaneously.

Within the cambial zone both periclinal and anticlinal divisions occur. The former extend throughout the cambial zone, but are usually most frequent in the zone of xylem mother cells. In conifers anticlinal divisions are virtually confined to the cambial initials, although they occur occasionally in the xylem mother cells and rarely in the phloem mother cells. Normally only about 2% of anticlinal divisions occur outside the initial layer, although the figure tends to be correlated with growth rate and may be higher in fast-growing trees. In dicotyledons anticlinal divisions may again be virtually restricted to a single cambial layer, as in *Leitneria floridana* Chapm. (Cumbie, 1967a), or they may be distributed over several layers, as in *Hibiscus lasiocarpus* (T. & G.) Gray (Cumbie, 1963) and *Acer pseudoplatanus* L. (Catesson, 1964).

Periclinal divisions lead to an increase in the amount of secondary vascular tissue. As the volume of secondary xylem increases, a tangential stress is produced on the vascular cambium immediately outside it, and this stress is countered by extension of the cambial circumference. Such extension might be brought about by increase in the tangential dimensions of the cambial initials, by elongation of the initials with intrusive growth so that a greater number of cells intersect a given transverse plane, or by anticlinal division of the initials. The relative importance of these possibilities was discussed by Bailey (1923). While in many plants the cambial initials show an increase in tangential and longitudinal dimensions with age, this is not sufficient to account for the amount of increase in cambial circumference. The tangential expansion of the cambium is largely the result of anticlinal divisions in the cambial initials followed by expansion of the daughter cells to the size of the parent. In storeyed cambia this takes the form of radial longitudinal division followed by tangential expansion; in non-storeyed cambia the divisions are obliquely transverse (pseudotransverse) and the daughter cells expand longitudinally.

Opinion remains divided on two important features of the cambium which will be discussed in this chapter. These are, first, the point in the ontogeny of the shoot at which the cambium appears and, second, the presence within the cambium of permanent initial cells.

Origin of the cambium

The precise time of origin of the cambium is difficult to define both theoretically and practically. There are two opposing and extreme viewpoints, though most morphologists have taken up some intermediate, usually undefined, position. On the one hand, the cambium may be regarded merely as a meristem producing regular radial files of derivative cells; if this definition is applied rigorously, in many plants much of the metaxylem must be regarded as a product of a cambium and therefore as a secondary tissue, although still undergoing extension growth. This view was advanced by Priestley (1928) and is responsible for the reports of fascicular cambial activity in many monocotyledons (Chapter 8). Primary tissues are often regarded as products of the apical meristem, but in fact most of the divisions producing them are located in the axis below the apex. Lateral growth is therefore a continuous process, unbroken from the apex to the mature trunk. It is useful to separate the earlier part of this process as primary growth and the latter as secondary growth, but few developmental anatomists have attempted to determine whether any clear transition can be observed and, if so, to define the precise point at which procambial divisions cease and cambial divisions begin. A study of the transition between procambium and cambium in *Canavalia* (Leguminosae) has been made recently by Cumbie (1967b). Early in its development the procambium becomes organized into two systems, one of elongated cells, the other of short cells in axial strands (Fig. 1.3). Cumbie points out several respects in which the cambium resembles the procambium. Vascular tissues are produced by periclinal divisions in both, though in the primary vascular tissues this is so only in the xylem. The elongated cells of the procambium

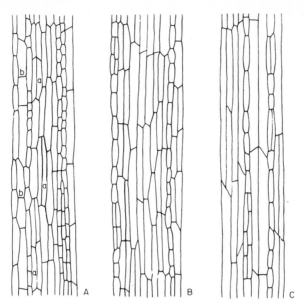

Fig. 1.3 Procambium of *Canavalia* in tangential view. The cells are organized into two systems, one of long cells, the other of short cells. A, early stage; B, later stage, C, end of primary growth. (From Cumbie, 1967.)

become the fusiform initials of the cambium, and the strands of shorter cells are converted into ray initials. In addition, some of the long procambial cells also form ray initials, just as these may arise by the subdivision of fusiform cells in the cambium. This close similarity in structure and development suggests that the recognition of two forms of the vascular meristem may not be justified in *Canavalia*. On this view the procambium and the cambium represent two developmental stages of the vascular meristem and it becomes meaningless to speak of the time of origin of the cambium.

On the other hand, Cumbie found that there were features in which the procambium and cambium differed in this genus. Most of the end walls of the elongated procambial cells are essentially transverse, whereas most of those of the fusiform initials are abruptly tapered. The cells of the procambium elongate during the active growth in length of the stem, whereas the length of the cambial initials remains relatively constant as in other storeyed cambia. Transverse divisions occur in the procambium, but were not observed in the cambium.

A greater contrast between the procambium and cambium in *Nicotiana* is illustrated by Esau (1965). The narrow radial dimensions indicate a high frequency of periclinal division in the cambium, and the two organized systems of fusiform and ray initials contrast with the more homogeneous procambium in that species. The transition from procambium to cambium was found by Catesson (1964) to be more abrupt in *Acer pseudoplatanus* than the situation described by Cumbie. In particular, the procambium was not differentiated into long and short cells as in *Canavalia*. Once elongation of the internodes ceases, procambial cells elongate by intrusive growth and acquire markedly pointed ends. These cells therefore become the fusiform initials. The onset of secondary growth is marked also by the conversion of some of these fusiform initials to ray initials. This is effected by a number of transverse divisions taking place in one fusiform initial resulting in a vertical file of ray initials. Rays also form in the interfascicular sectors of the stem. In *Acer* the medullary rays are only two or three cells wide. When primary growth ceases these interfascicular sectors are formed of slowly differentiating parenchyma. They now begin to divide tangentially, functioning as ray initials, of which they have all the characteristics. As secondary growth proceeds, the large rays of the interfascicular sectors become dissected into smaller rays by the intrusion of fusiform initials.

Catesson emphasizes that she does not consider the cambium to be merely a residue, a later phase, of the procambium in which divisions have become concentrated and regular. She regards it as a distinctive tissue, which in the mature state is characterized by a differentiation into fusiform and ray initials, and an increase in length of the former as intrusive growth begins, together with an accentuation of the acute ends of these cells. There is much in favour of Catesson's concept of the cambium, but most authors do not take so extreme a view. In many plants young internodes still actively increasing in length may contain bundles which are expanding rapidly by a meristem which has every appearance of a cambium, *at least in transverse view*, and is in fact continuous with the undoubted cambium in adjacent more mature internodes. These meristems are commonly regarded as cambial, and this is the position adopted by Sterling (1946) in his work on *Sequoia sempervirens* (Lamb.) Endl. The early ontogeny of the cambium must become known in several more plants before this fundamental problem can be resolved. It is likely that great diversity will be revealed.

The complete cambial cylinder may develop in a variety of ways. If the metaxylem forms a complete cylinder before secondary growth begins, the cambium will be complete from the time of its inception. More frequently, the primary bundles are distinct at the time of cambial initiation: in that event regular divisions usually occur earlier in the bundles than in the interfascicular sectors. Indeed, at any particular level of the stem, regular divisions, whether strictly cambial or not, will begin in the larger, older bundles of a twig before they occur in those formed later. Additional bundles may be interpolated after cambial activity has begun. Such bundles are called intermediate bundles by de Bary (1884) and are, of course, composed wholly of

secondary tissues. In most woody plants the interfascicular sectors of the
cambium are completed very soon after the fascicular portions. Beijer (1927)
and Catesson (1964) have described the extension of typical cambial cells
from the fascicular into the interfascicular sectors. Once the cambial cylinder
is complete, the secondary tissues it produces are usually uniform over its
whole circumference. Plants in which the interfascicular sectors produce
distinctive secondary xylem are discussed in Chapter 6.

Although the cambium of roots is similar in structure and action to that of
stems, its origin is necessarily different because of the characteristic arrangement of primary tissues. The initiation of the cambium in the roots of *Pyrus*
has been described by Esau (1943), and this example is typical of roots
generally. The cambium first appears as arcs of regular divisions situated to
the inner side of the phloem groups. These arcs subsequently become united
into a complete, but fluted, cylinder by divisions occurring in the pericycle
opposite to the protoxylem. The cambium becomes circular in section by the
early production of xylem derivatives opposite the phloem.

The permanence of the initials

Opinion still remains divided on the important question of the nature of the
cambial initials. Three views have been proposed. The first is mainly of
historical interest. Hartig (1853) considered that the cambium was a cylinder
two cells thick. The outer layer consisted of initials from which the phloem
was derived and the inner layer of initials giving rise to the xylem. These
initials stood back to back in pairs so that the files in the phloem corresponded to those in the xylem. Sanio (1873) rejected this concept. He argued
that, since at mitotic cell division a new membrane is deposited around each
daughter protoplast (in addition to the partition wall), the tangential wall
between two such initials would become thicker and thicker with successive
cell divisions until in an old tree-trunk it would be several millimetres thick.
Such a thick wall has never been observed in the cambium. It is true that thick
walls do occur occasionally in the cambial zone, but this is thought to be due
to a single initial cutting off a number of cells in one direction before
reversing to cut off cells in the opposite direction. If for several successive
divisions the inner daughter cell of the initial becomes a xylem mother cell
while the outer one retains the function of initial, the tangential wall on the
phloem side of the initial will become increasingly thick due to the deposition
of successive walls around the new protoplast after each division. If an
exceptionally large number of successive derivatives are cut off towards the
xylem this wall will become abnormally thick. When the initial begins to cut
off cells in the opposite direction the thick wall will be carried into the phloem
mother-cell zone and eventually into the mature phloem. Similarly, exceptionally thick walls may be carried into the xylem, although this is less common,

as phloem is produced in smaller quantities than xylem. If a new membrane is deposited around the protoplast of each new daughter cell it is apparent that the radial walls of an initial will be added to continually. Sanio noted this point and observed that the radial walls of cells in the cambial zone were in fact very thick, even though subjected to continual radial tension. Increase in radial wall thickness is presumably balanced by radial expansion of the initial after each division.

Sanio pointed out a second objection to Hartig's concept of double initials. The similar arrangement of the files of cells in the phloem and xylem on opposite sides of the cambium supports the concept of an initial common to xylem and phloem. This point is made also by Schoute (1902), and Bannan (1955) describes the simultaneous origin or loss of radial files in the xylem and phloem.

The second concept, which postulates a multiseriate cambium in which all cells are equivalent, was put forward by Raatz (1892), was adopted by Kleinmann (1923), and has recently been given strong support by Catesson (1964). She considers that the active cambium is formed of several layers of similar cells, each endowed with equivalent powers of multiplication. She regards all as initial cells. The cambium, therefore, like meristems in the apex,

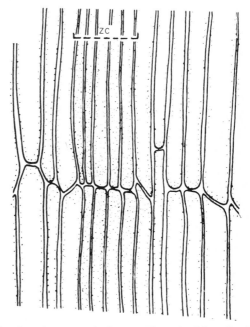

Fig. 1.4 Radial section of the cambial region of *Acer pseudoplatanus*. The ends of the cells of the cambial zone (ZC) are at approximately the same level. (From Catesson, 1967.)

consists of a mass of cells within which the initiating role is not reserved to a privileged layer. She supports this opinion with many observations. In spite of meticulous examination of numerous sections of the active cambium of *Acer pseudoplatanus*, she found it quite impossible to distinguish a layer in which divisions were most frequent. It was not that the cambial initial layer divided less often than its derivatives, as Bannan (1955) found in *Thuja occidentalis* L. In *Acer*, mitoses were found to be abundant in all the cells in a zone several layers thick, and these cells all shared the characteristics of truly meristematic cells. Moreover, the terminations of these cells, in radial view, lay almost exactly at the same level (Fig. 1.4), in contrast to the situation in *Thuja* (Fig. 1.5). Nor was it possible to select a unique initial for each file of cells by using other criteria, whether histochemical, anatomical, or cytological. Moreover, she did not consider that the pattern of periclinal divisions supported the idea of a single initial layer, because contiguous cells of the same file were frequently seen in the same stage of active mitosis.

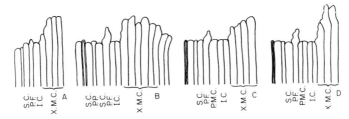

Fig. 1.5 Drawings from radial sections of the cambial region of *Thuja occidentalis*, showing differences in the length of the various types of cell. SC, sieve cell; PP, phloem parenchyma; PF, phloem fibre; IC, initial cell; PMC, phloem mother cell; XMC, xylem mother cell. (From Bannan, 1955.)

This work of Catesson is the most serious criticism of the third view of the cambium, which nevertheless is that most widely held at the present time. It postulates a single permanent initial cell in each radial file of the cambial zone. Such an initial lies between phloem and xylem mother cells. When it divides periclinally it adds a cell either to the outside (new phloem mother cell) or to the inside (new xylem mother cell), but in either event one daughter cell remains as the continuing initial. Evidence in favour of a single permanent initial has been provided by Bannan (1955, 1968) and Newman (1956), who both succeeded in distinguishing a single initial cell between the phloem and xylem mother cells. Bannan (1955) found that, in *Thuja*, the three constituent layers of the cambium could be distinguished in radial sections by the appearance of their tips (Fig. 1.5). At the inside of the phloem, in each radial file of elements, there are usually one to four short cells, which are about the same length as the most recently differentiated sieve cell or phloem parenchyma. These are the initial and the phloem mother cells or their immediate derivatives. The xylem mother cells are somewhat longer. New-man, using transverse sections, deduced the position of the cambial initials in very wide cambial zones in *Pinus radiata.* He identified the initial region by locating the cells of the initial region of the rays, which may be distinguished from maturing ray cells by their dense cytoplasm and short radial dimension, and by observing very recent radial divisions, which are virtually restricted to the cambial initials. He then located the cambial initials with some certainty using characters of partitioning and wall thickness. Sanio (1873) had shown that in *Pinus sylvestris* L. each cambial derivative divided once to give two cells destined to differentiate as either xylem or phloem. These pairs of cells were recognizable in the files of the cambium by characters of wall thickness and the type of abutment of the tangential on to the radial walls. The last-formed pair could be seen to form a group of four cells comprising two destined to be vascular elements and two which represent the initial and its daughter cell. This daughter cell would in turn produce a pair, and so the process would repeat itself to one side or other of the initial. It is now known that in more rapidly dividing cambia of *Pinus* the vascular mother cells may be grouped in pairs (or sometimes threes), but Newman still found it possible to distinguish, though not always with complete certainty, the group of cells which included the initial. Further work by Mahmood (1966) in Newman's laboratory has produced electron-microscope evidence for the wall characters upon which Newman's judgements rested.

In conifers, anticlinal divisions are virtually restricted to the cambial ini-tials, as evidenced by the presence of the resultant doubled radial files of cells in the vascular tissue to both sides of the cambium. Other features also correspond in the xylem and the phloem, including the loss of radial files of axial cells and the replacement of such files by rays. These are caused by, respectively, the loss of fusiform initials from the cambium and the conver-sion of fusiform initials to ray initials. If the initiating layer were more than one cell wide an exact duplication of all anticlinal divisions and loss and conversion of fusiform initials would be required in the several initials of a radial file, in order to produce corresponding patterns in xylem and phloem. In view of the complex and variable behaviour of fusiform initials in a conifer cambium, such duplication is manifestly unlikely.

It seems clear that in conifers the cell pattern in the xylem and phloem is determined by a single cambial initial in each radial file. It may well be significant, however, that the evidence for a single, permanent initial is derived from conifers, while the chief recent dissenter from this view worked with a dicotyledon. Cambial behaviour is complex and fairly uniform throughout the conifers. In dicotyledons, on the other hand, it varies from a condition similar to that in conifers to a much simpler one with anticlinal divisions of the radial longitudinal type, no loss of fusiform initials and conversion of fusiform to ray initials by simple transverse segmentation. The small amount of data available for the distribution of anticlinal divisions in dicotyledons shows that they may be either virtually restricted to a single layer (Cumbie, 1967a) or distributed over several layers (Catesson, 1964; Cumbie, 1963). The width of the initiating layer may prove to be similarly variable; certainly existing evidence is insufficient to indicate that a single initiating layer is invariable or even general in dicotyledons.

REFERENCES

BAILEY, I. W. (1920). The cambium and its derivative tissues. II. Size variations of cambial initials in gymnosperms and angiosperms. *Am. J. Botany* **7,** 355–67.
— (1923). The cambium and its derivative tissues. IV. The increase in girth of the cambium. *Am. J. Botany* **10,** 499–503.
BANNAN, M. W. (1955). The vascular cambium and radial growth in *Thuja occidentalis* L. *Can. J. Botany* **33,** 113–38.
— (1957). The relative frequency of the different types of anticlinal divisions in conifer cambium. *Can. J. Botany* **35,** 875–84.
— (1968). Anticlinal divisions and the organization of conifer cambium. *Botan. Gaz.* **129,** 107–13.
BEIJER, J. J. (1927). Die Vermehrung der radialen Reihen im Cambium. *Rec. Trav. Bot. neerl.* **24,** 631–786.
CATESSON, A. M. (1964). Origine, fonctionnement et variations cytologiques saisonnières du cambium de l'*Acer pseudoplatanus* L. (Acéracées). *Ann. Sci. nat. (Bot.) 12e ser.* **5,** 229–498.
CUMBIE, B. G. (1963). The vascular cambium and xylem development in *Hibiscus lasiocarpus. Am. J. Botany* **50,** 944–51.
— (1967a). Developmental changes in the vascular cambium of *Leitneria floridana. Am. J. Botany* **54,** 414–24.
— (1967b). Development and structure of the xylem in *Canavalia* (Leguminosae). *Bull. Torrey Bot. Club* **94,** 162–75.
DE BARY, A. (1884). *Comparative Anatomy of the Vegetative Organs of Phanerogams and Ferns,* Oxford University Press, Oxford.
ESAU, K. (1943). Vascular differentiation in the pear root. *Hilgardia,* **15,** 299–324.
—(1965), *Vascular Differentiation in Plants,* Holt, Rinhart and Winston, New York.
GREW, N. (1682). *The Anatomy of Plants,* 2nd ed., Rawlins, London.
HARTIG, T. (1853). Ueber die Entwicklung des jahrringes der Holzpflanzen. *Bot. Zeit.* **11,** 533–66, 569–79.
KLEINMANN, A. (1923). Ueber Kernund Zeilteilungen im Cambium. *Bot. Arch.* **4,** 113–47.
KRIBS, D. A. (1935). Salient lines of structural specialization in the wood rays of dicotyledons. *Botan. Gaz.* **96,** 547–57.

MAHMOOD, A. (1968). Cell grouping and primary wall generations in the cambial zone, xylem, and phloem in *Pinus. Australian J. Bot.* **16,** 177–96.

NEWMAN, I. V. (1956). Pattern in meristems of vascular plants. 1. Cell partition in living apices and in the cambial zone in relation to the concepts of initial cells and apical cells. *Phytomorphology* **6,** 1–19.

PRIESTLEY, J. H. (1928). The meristematic tissues of the plant. *Biol. Rev.* **3,** 1–20.

RAATZ, W. (1892). Die Stabbildungen im secondären Holzkörper der Bäume und die Initialentheorie. *Jahrb. wiss. Bot.* **23,** 567–636.

SANIO, K. (1863). Vergleichende Untersuchungen über die Zusammensetzung des Holzkörpers. IV. *Bot. Zeit.* **21,** 401–12.

— (1873). Anatomie der gemeinen Kiefer (*Pinus sylvestris* L.). *Jahrb. wiss. Bot.* **9,** 50–126.

SCHOUTE, J. C. (1902). Ueber Zellteilungsvorgängerim Cambium. *Verhoudel. Akad. Wetenschappen, 2s.* **9,** 1–60.

STERLING, C. (1946). Growth and vascular development in the shoot apex of *Sequoia sempervirens* (Lamb.) Endl. III. Cytological aspects of vascularisation. *Am. J. Botany* **33,** 35–45.

2
Cell structure and growth cycles

The cells of the cambial zone are of two types, the elongated fusiform initials and the nearly isodiametric ray initials. The fusiform initials are elongated in an axial direction, usually being several hundred times as long as their radial diameter (Bailey, 1919). They tend to be tangentially flattened and pointed at both ends. The exact shape of the fusiform initials of *Pinus sylvestris* has been determined by Dodd (1948), who found that the initials had from 8 to 32 faces with an average of 18 faces. Each fusiform initial was found to be in contact with an average of 14 other initials like itself. Fusiform initials illustrate considerable variation in their dimensions (Bailey, 1920b). Variation occurs between different species of plants; for example, in *Robinia*, a plant with a storeyed cambium, the fusiform initials are about 175 μm long compared to *Pinus*, a conifer with a non-storeyed cambium, where the fusiform initials are about 3,500 μm long. Further variation in cell dimensions occurs within actual plants (Chapter 4).

While our knowledge concerning most other aspects of the vascular cambium has increased more or less continuously over the years, the cell cytology of the cambium has remained virtually unexplored from the pioneering efforts of Bailey (1919, 1920a, 1920c) until the present decade. This no doubt has been largely due to the technical difficulties encountered in attempting to study the actual cells of the cambial zone while they are still in the living state. The advent of modern techniques for isolating and preserving cells combined with the use of the electron microscope has now opened the way for renewed interest in this fundamental field.

Bailey (1920a) showed that, despite their comparatively large size, the fusiform initials in the gymnosperms studied were uninucleate. Each initial contained a single nucleus centrally located, retaining its position during 'karyokinesis'. Furthermore, he found, despite very wide variations in cell length, that the size of the nucleus did not tend to increase with increase in cell length, nor was there any tendency to polyploidy. Bailey also described the nuclear changes and the formation of the cell plate during the extraordinary periclinal divisions found in the cambial fusiform initials.

In more recent years a number of papers by various authors have described the ultrastructure of the vascular tissues of certain plants, but usually they have been more concerned with the differentiation of the xylem or phloem (e.g. Cronshaw and Wardrop, 1964; Esau, Cheadle and Gill, 1966) than with the cambium itself. Hohl (1960) made a study of the differentiation of cell organelles in the course of normal histogenesis in *Datura stramonium* L. His observations suggest that the cambial cells are very similar in ultrastructure to parenchyma cells.

The most detailed contributions so far to our knowledge of cambial ultrastructure have been the work of Srivastava and O'Brien. Srivastava (1966) reported on the cambium of an angiosperm, *Fraxinus americana* L. Basically the fusiform and ray initials show the same structure. Both are highly vacuolate, with only a thin layer of cytoplasm containing the usual complement of organelles and membranes typical of parenchyma cells. Actively dividing cambial cells are highly vacuolate, rich in endoplasmic reticulum of the rough cisternal form, ribosomes, golgi bodies and coated vesicles. Microtubules were found to be abundant in the peripheral cytoplasm. The nucleus is large, with a single nucleolus, the nuclear envelope having well-defined pores. The quiescent cambium of *Fraxinus* was found to be similar to the summer material, though the rough cisternal form of the endoplasmic reticulum and the coated vesicles were not so abundant. An earlier study on *Pinus strobus* L. (Srivastava and O'Brien, 1966) disclosed a similar fine structure to that in *Fraxinus*, the gymnosperm and angiosperm differing only in detail.

Robards and Kidwai (1969) studied the seasonal variations in the ultrastructure of the resting and active cambium of *Salix fragilis*. Their findings in general confirm those of Srivastava, but point to the large amount of space occupied by protein bodies, lipid droplets, and vesiculate smooth endoplasmic reticulum as a striking feature of the resting cambium. They suggest that these bodies are storage materials which are required during the first stages of differentiation at the beginning of the growing period.

The organization of the cambial cell wall has been the subject of a number of studies over the years. The conifer cambium has been examined by optical means (Kerr and Bailey, 1934), X-rays (Preston and Wardrop, 1949), and the electron microscope (Hodge and Wardrop, 1950; Wardrop, 1954; Preston and Ripley, 1954). These studies have helped to establish the existence of a dispersed wall texture in which the orientation of the microfibrils in the outer layers is predominantly transverse. Chemically, cambial cell

walls are built up of the usual cellulose associated with non-cellulosic substances. They have primary pit fields with plasmodesmata. The radial walls
are generally thicker than the tangential walls, especially during dormancy,
and their primary pit fields are deeply depressed.

Seasonal variations in the organization of the cambial cell have been
recorded. Bailey (1930) studied the vacuome in fusiform initials of a number
of plants and showed important seasonal changes. Numerous small vacuoles
exist in winter material, but there is generally only one large vacuole per cell
in summer. These transformations are apparently reversible, and transition
forms are numerous. Catesson (1962) has also reported on the changes in
state of the vacuoles in passing from summer to winter conditions.

Buvat (1956) has suggested that the mitochondria in cambial initials of
Robinia are found in chains in winter but occur singly in summer. Further
studies on seasonal changes within the cambium have been published by
Catesson (1961, 1964). Thimann and Kaufman (1958) studied the cytoplasmic streaming in cambial cells of *Pinus*. They recorded a maximum velocity
in the spring, with a subsequent decline in the average rate during the summer. Priestley (1930b) observed that the protoplasm in hardened cambial
cells was densely granular and suggested that the protoplasm was in a gel
state in winter and a sol state in spring. The cell contents are swelling at this
stage, and it is more difficult to observe unstained tissue. The cell walls, which
were thick in winter, become thinner in spring. In the pine Srivastava and
O'Brien (1966) observed marked changes in the state of vacuoles, plasmalemma and endoplasmic reticulum in the transition of the cambium from
winter to summer condition. In the winter the vacuoles were small and
numerous, the plasmalemma was thrown into folds, and the endoplasmic
reticulum occurred mostly in the form of smooth vesicles. These were
replaced in summer by one or two large vacuoles, a more smooth and regular
plasmalemma, and the rough cisternal form of the endoplasmic reticulum.
Fine cytoplasmic fibrils were also seen in the winter material. While it is
recognized that it is not possible to distinguish between the cambial 'initial'
and those derivatives in front and behind it in the cambial zone with normal
light-microscope observations, it is worth noting that even at the ultrastructure level these authors could not identify a distinct initial. They noted that
this difficulty in distinguishing the initial from the derivative held for the ray
initials also, but to a lesser extent.

The initials of the cambium divide in two directions and in a number of
different ways. The phloem and the xylem are formed by periclinal
(tangential) divisions. These vascular tissues are laid down in two opposite
directions, normally the xylem to the inside and the phloem to the periphery
of the axis. Normal periclinal division is brought about by the formation of a
vertical cell plate running the length of the cell in the tangential plane. Such a

division produces two radially adjacent daughter cells, one of which assumes the function of the initial, while the other becomes a xylem or phloem mother cell. Anticlinal division in the fusiform initials is of two main types. Radial longitudinal division occurs in storeyed cambia through the formation of a vertical cell plate from one end of the cell to the other in a radial plane. Pseudotransverse division is characteristic of non-storeyed cambia. Here the cell plate is obliquely transverse, intersecting the two radial walls at different levels, and after division the daughter cells undergo elongation. Anticlinal divisions lead to an increase in the circumference of the cambium, through the tangential expansion of the daughter cells in the storeyed cambia and their expansion and elongation in non-storeyed cambia.

Periclinal divisions

Periclinal division in fusiform initials has been described by Bailey (1919, 1920a, 1920c). The process of cell-plate formation is greatly extended in time, since the plate is laid down along the length of a very elongated cell. The nucleus, which is located in about the centre of the cell, divides mitotically, and the spindle becomes extended laterally by the addition of peripheral fibres, gradually assuming the form of a disc. The original 'connecting fibres' disappear from about the cell plate as new kinoplasmic fibres are successively added to the growing cell plate. This ring increases in circumference until it intersects the radial walls, when it becomes more or less flattened on two sides. As soon as the cell plate intersects the radial walls of the cell, the fibres in these two sides disappear, leaving two separate aggregations of kinoplasmic fibres which are connected by the cell plate. These aggregations of fibres, or kinoplasmasomes, extend across the cell, more or less at right angles to its long axis from one radial wall to the other, and are located in the centre of the

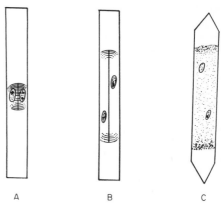

A   B   C

Fig. 2.1 Diagrams showing stages in periclinal division of a fusiform initial. A, B, early and later stages of division as seen in radial section; C, as seen in tangential section.

protoplast, midway between its tangential surfaces. They move in opposite directions towards the ends of the cell, and as they do so the cell plate is extended until it reaches the ends of the protoplast, thus dividing it into two halves, each containing one of the daughter nuclei (Fig. 2.1).

Following periclinal division, one of the daughter cells remains in the initial layer, where it expands radially before dividing again. The other daughter cell becomes a xylem or phloem mother cell. It expands radially and may or may not divide again in the same manner before differentiating into a mature vascular element. Both Newman (1956) and Bannan (1950, 1955) show that mitosis is most frequent some distance from the initial cells. This is particularly true of the xylem mother cells, which both authors find to divide more frequently than those of the phloem. Both also compare this situation with the well-known fact that mitotic figures are more frequent in the peripheral parts of the shoot apex than in the central initial cells (Edgar, 1961).

There is some difference of opinion regarding the frequency of redivision of the mother cells on either side of the initials. This is mainly due to real differences between the species investigated and to differences resulting from environmental causes. Sanio (1873) considered that the xylem mother cells of *Pinus sylvestris* divide once or occasionally twice before maturation. Newman (1956) deduced from cell grouping in transverse sections of *P. radiata* that the xylem mother cells divide twice, so that the elements of the xylem tend to occur in groups of four. Other workers, particularly Raatz (1892) and Wilson (1964), have considered that the mother cells undergo several periclinal divisions, one cell being capable of producing more than twenty derivatives. Bannan (1957a, 1962) has discussed this problem. Several lines of evidence support the view that xylem mother cells divide to give several derivatives. The grouping of cells as seen in transverse section indicates the number of cell generations. In radial longitudinal section cell tips are associated in groups, which can be interpreted as the progeny of a single mother cell. As many as eight tips may be found in a group, showing that divisions of at least the third order have occurred. In broad zones of periclinal divisions, which may be thirty meristematic cells in breadth, it is found that the maximum frequency of division is at the central part of the zone, indicating that certain xylem mother cells are actively redividing. Also, the occurrence of doubling of the files of xylem elements over a limited distance suggests that a mother cell has undergone anticlinal division, and that the two daughter cells have continued to redivide periclinally. The fact that these doubled files sometimes extend over a considerable distance suggests that the products of the divided cell may be numerous.

The further development of the mother cells into xylem and phloem elements takes place outside the cambium. Recent information on the differentiation of vascular tissues can be found in the books edited by Zimmermann (1964) and Côté (1965).

Wilson (1964) estimated that the length of the division cycle, including division, synthesis, and enlargement, is about ten days in *Pinus strobus*. The process of division occupies one day, with mitosis taking five hours and phragmoplast movement the remaining nineteen hours. The approximate times occupied by the individual mitotic phases are estimated to be: prophase, 2·4 hrs; metaphase, 0·4 hr; anaphase, 0·7 hr; and telophase, 1·6 hr. The overall rate of phragmoplast movement is calculated to be 50–100 μm/hr.

Wilson suggested that possibly cell division in a population, once initiated, occurs at a maximum rate limited by the minimum time required for the cycle. During the grand period of growth the frequency of division may be limited by the time involved in the separate and sequential reactions of the growth process rather than the availability of materials, energy sources, or hormones.

The length of the cell cycle in *Pinus strobus*, based on the mitotic index, is at a minimum early in the season, and increases throughout the season until it is more than twice its minimum value (Wilson, 1966). The mitotic index may double (thus doubling the rate of cell production) without any increase in the number of cells in the cambial zone. A similar effect has been recorded for *Picea glauca* (Moench) Voss (Gregory and Wilson, 1968). Trees of comparable size in Alaska, although producing derivatives twice as fast as those growing in New England, had the same number of cells in the cambial zone, but the mitotic index was doubled.

There are few studies on the distribution of the mitoses across the cambium. Bannan (1955) and Wilson (1964) both record a peak near the centre of the cambial zone, though Bannan has observed that the first mitoses in the spring occur near the xylem. Since the ratio of phloem to xylem production is lower in the spring than the autumn, it has been suggested by Wilson and Howard (1968) that this might be interpreted as a shift in the zone of peak division from the xylem to the phloem side in autumn. Wilson and Howard (1968) have used a computer model to study aspects of the cambium in *Pinus strobus*. From the results they suggest that each radial file of cambial cells must have a different distribution of mitotic frequency. The model was also used to study the number of times each xylem mother cell redivides, most mother cells redividing at least three times.

Anticlinal divisions

Nägeli (1864) was one of the first to speculate on the mode of division by which new cambial cells were formed. He proposed a theoretical scheme in which the cambial cells divided by radial longitudinal walls. Some workers, however, did not accept this generalization. Hartig (1895) inferred from the structure of the secondary xylem in *Pinus sylvestris* that in conifers the fusiform initials divided transversely and then elongated to increase the cambial circumference. Klinken (1914) reached a similar conclusion from the study of serial sections of the phloem of *Taxus baccata* L. He considered that there were two fundamental types of meristematic activity: the first involved radial longitudinal division of fusiform initials and was characteristic of dicotyledons; in the second, characteristic of the conifers, the divisions were transverse and the fusiform initials elongated to increase the cambial girth. Neeff (1920) found that the type of cambial activity ascribed to the conifers by Hartig and Klinken was present in the xylem and phloem of a dicotyledon, *Tilia tomentosa* Moench. This led him to conclude that no fundamental distinction existed between the cambia of gymnosperms and dicotyledons.

Bailey (1923) pointed out that it was not possible to ascribe particular types of meristematic activity with certainty to large groups of vascular plants from the study of one or two supposedly representative species. He secured material of the cambium from a wide series of gymnosperms and angiosperms of both temperate and tropical regions, and found a strict correlation between the arrangement of fusiform initials and their method of division. In the gymnosperms and the less structurally specialized dicotyledons the fusiform initials are not arranged in regular tiers, and the cambium is said to be non-storeyed (Fig. 1.2). In certain of the more highly specialized dicotyledons the fusiform initials are arranged in tiers and the cambium is said to be storeyed (Fig. 5.1).

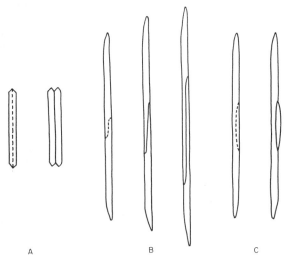

A   B   C

Fig. 2.2 Diagrammatic representation of the types of anticlinal division in fusiform initials. A, radial longitudinal; B, pseudotransverse; C, lateral.

In storeyed cambia it is clear that the increase in circumference is not due to the elongation of transversely dividing cells, since in such circumstances the newly formed initials would crowd past one another and the stratified arrangement would be lost. Division is in fact by radial longitudinal walls (Fig. 2.2A). This results in a grouping of the initials in horizontal rows, provided that differential elongation of the daughter cells does not occur. It is equally clear that if the irregular arrangement of non-storeyed cambia were due to the differential elongation of the products of radial longitudinal divisions the average length of initials would increase considerably with each division. This does not occur. Cell multiplication in these non-storeyed cambia is by more or less horizontal (pseudotransverse) dividing walls, with a subsequent increase in length of the daughter cells (Fig. 2.2B).

The stratified cambium and radial longitudinal type of anticlinal division in structurally highly evolved dicotyledons was considered by Bailey to be a development from the non-storeyed cambium with pseudotransverse anticlinal division. As the structure becomes more specialized, the fusiform initials become shorter and the ends of the oblique partitions formed in anticlinal division approach the ends of the initials, until eventually the partition reaches from one end of the initial to the other, and becomes radial longitudinal. An intermediate state may exist, as in the herbaceous dicotyledon *Hibiscus lasiocarpus*, in the cambium of which radial longitudinal and pseudotransverse divisions occur with almost equal frequency. Following the radial longitudinal divisions there is little elongation of the fusiform initials, while elongation subsequent to pseudotransverse division is variable but insufficient to account for much increase in circumference of the cambium (Cumbie, 1963).

In pseudotransverse division the dividing wall is usually laid down near the centre of the cell; less commonly it may be nearer to one end, resulting in daughter cells of unequal size. The length and pitch of the wall are usually very variable, ranging from short and transverse or nearly so to very oblique and up to half the length of the dividing cell in conifers (Bannan, 1957b) and considerably more in dicotyledons (Evert, 1961; Cumbie, 1967, 1969). Exceptions to this tendency have been observed in *Larix europea* DC., in which the walls are always strongly oblique (Hejnowicz, 1961), and in *Ginkgo biloba* L. (Srivastava, 1963b) and *Acer pseudoplatanus* (Catesson, 1964), in which positions approaching the transverse are rather rare. In conifers the length of the wall is related to the length of the dividing cell, with long cells tending to have long partitions, and vice versa (Bannan, 1964a, 1965). In the dicotyledon *Leitneria floridana* the wall is shorter in old than in young material, decreasing from an average of 50% of the mother cell in the first year of secondary growth to 24% in years 22–26. The wall is longer in relation to the mother-cell length in *Leitneria* than in *Pyrus communis* L., and this may be related to the fact that *Leitneria* has shorter fusiform initials (Cumbie, 1967).

In conifers the direction of slope of the dividing wall tends to be the same in neighbouring cells of the cambium, although exceptions occur. After varying lengths of time, reversals in tilt take place; these are not necessarily synchronized over the entire cambial area, and different sectors of the cambium may have partitions sloping in different directions. Continued anticlinal divisions in one direction lead to the development of spiral grain in the wood, and this process is accelerated by a high frequency of anticlinal division. Periodic reversals prevent the production of a pronounced spiral grain, and the interval between reversals, although highly variable, tends to be inversely proportional to the frequency of anticlinal division (Bannan, 1963a, 1963b, 1964a, 1964b, 1965, 1966a, 1966b; Hejnowicz, 1961, 1964). Hejnowicz (1964) has proposed the term *domain* to designate a sector of the cambium where the anticlinal partitions are tilted in one direction. Domains differ in size and the domain pattern changes with time. Little is known about the direction of slope of the partition in dicotyledons, but in *Leitneria* it slopes about equally in both directions in a given sector, and often reverses from mother to daughter cell (Cumbie, 1967).

In most cambia anticlinal divisions occur which are neither pseudotransverse nor centrally radial longitudinal. The dividing wall is laid down longitudinally and to one side of the fusiform cell and opposite the centre. It does not extend the full length of the cell and results in a cell being cut off the side of the fusiform initial (Fig. 2.2C). Such divisions, which intercept only one lateral wall of the fusiform initial, have until recently been termed (*longitudinal*) *divisions off the side*. Cheadle and Esau (1964) have called them *lateral divisions*, and this terminology is adopted here. Barghoorn (1940a, 1940b) and Braun (1955) have described such divisions as a common method of ray origin in conifers and dicotyledons, but other workers have found them to be rare in gymnosperm stems (Bannan, 1950, 1951a, 1957b, 1965, 1968a; Whalley, 1950; Hejnowicz, 1961; Srivastava, 1963b). Bannan (1968a), studying a large number of conifer species, found that the overall frequency of lateral divisions is less than 1% of the total of anticlinal divisions. The limited data available for dicotyledons do not permit any definite conclusions to be drawn, although they suggest that lateral divisions tend to be more frequent than in conifers and that comparative frequency within the dicotyledons is related to the length of the fusiform initials. In *Pyrus communis* (Evert, 1961) and *Liriodendron tulipifera* L. (Cheadle and Esau, 1964) about 5% of anticlinal divisions are lateral. In *Leitneria floridana*, which has shorter initials, less than 5% of divisions in old material are lateral but up to 50% in the first year of secondary growth. This last figure may be compared with the situation in the herbaceous perennial *Hibiscus lasiocarpus*, which has still shorter initials. Here the proportion of lateral divisions is smaller, but 54% of anticlinal divisions are either lateral or radial longitudinal, the presence of radial longitudinal divisions probably being related to the shortness of the initials (Cumbie, 1967). In the herbaceous species *Polygonum lapathifolium* L., in which anticlinal divisions are either pseudotransverse or lateral, Cumbie (1969) found that almost 50% of the total number were lateral. The percentage of lateral divisions varied among different types of radial files, being 32% in interfascicular files containing only fibres, about 50% in interfascicular files containing both fibres and vessel elements, and approximately 60% in fascicular files.

The length of the partition in lateral divisions varies widely in all species which have been studied, being up to $\frac{1}{3}$ of the length of the fusiform initial in conifers, from $\frac{1}{4}$ to $\frac{2}{3}$ in *Hibiscus*, from 25 to 70% in *Leitneria* and from 15 to 94% in *Polygonum*.

It is only in recent years that high frequencies of anticlinal division in cambial initials have been demonstrated. Priestley (1930a) calculated from the data of Bailey (1923) that no more than four such divisions per cell would be required to produce the number of cells present after 60-yr growth of a pine stem. However, no data were published on the actual frequency of anticlinal divisions, except some limited observations by Klinken (1914) on a single piece of *Taxus* stem, until the papers of Bannan (1950) and Whalley (1950) on the conifers *Chamaecyparis* and *Thuja* respectively. Both these workers found that the frequency of anticlinal division in the fusiform cambial initials was far higher than once every 15 years, as estimated by Priestley, or once every 5 years, as noted by Klinken in *Taxus*. Evidence of active anticlinal division was observed in all parts of the tree, with an initial undergoing as many as three or four successive divisions during a single year's growth in fast-growing young stems. Such high frequencies of anticlinal division produce far more new fusiform initials than are necessary for the required increase in circumference of the cambium. In consequence, production of new fusiform initials is accompanied by their rapid elimination, so that only a slight increment results (Fig. 2.3).

Similar cambial behaviour, with high rates of production of fusiform initials and almost equally high rates of loss, has been observed repeatedly in conifers by Bannan and by other workers in both gymnosperms (Hejnowicz, 1961; Srivastava, 1963a, 1963b; Hejnowicz and Branski, 1966) and dicotyledons (Evert, 1961; Cheadle and Esau, 1964). However, in the herbaceous dicotyledons *Hibiscus lasiocarpus* and *Polygonum lapathifolium*, Cumbie (1963, 1969) observed no outright loss of fusiform initials from the cambium, and only very occasional loss following transformation to ray initials. During the formation of the inner secondary xylem in *Leitneria floridana* the production of new fusiform initials proceeds at a rate just about adequate for the circumferential expansion of the cambium and almost all initials survive. The rate is somewhat in excess during the later stages of growth, and during this period about 20% of the initials studied were lost from the cambium (Cumbie, 1967).

The frequency with which fusiform initials undergo anticlinal division varies greatly. It may be influenced by inherited genetic differences, as seen in

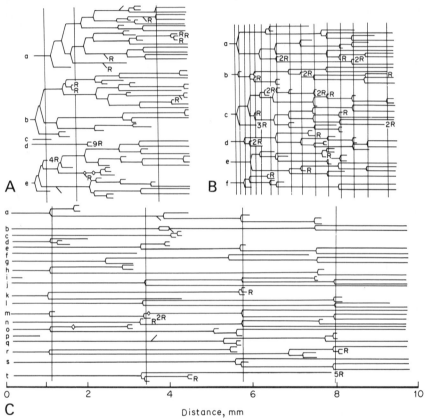

Fig. 2.3 Diagrammatic representation of the succession of tracheid rows in selected lineal series. The diagrams are to be interpreted as showing the relative time and frequency of anticlinal divisions involved in the production of new fusiform initials, the disappearance of fusiform initials from the cambium and the origin of rays. A, young, vigorous stem, *Chamaecyparis lawsoniana* (A. Murr.) Parl.; B, young, slow-growing stem, *C. thyoides* (L.) B.S.P; C, old vigorous stem 22 cm in dia., *C. thyoides*. Equal forkings of the horizontal lines indicate centred pseudotransverse division, unequal forkings, eccentric pseudotransverse division and side branches, lateral divisions. The termination of horizontal lines indicates the disappearance of fusiform initials from the cambium, and the letter R signifies the establishment of one or such a number of rays as indicated by the preceding digit. The vertical lines represent the boundaries of annual rings. (From Bannan, 1960b.)

the differing frequencies in different species or races, by factors in the external environment, as indicated by differing frequencies in trees from different sites and in wide and narrow growth increments in the same tree, and by conditions within the cambium, where neighbouring sectors and even adjacent initials may behave quite differently. It is not uncommon for sister initials descended from the same parent cell to show markedly different frequencies of division, but the reasons for such behaviour are unknown.

Factors which have been shown to affect the frequency of anticlinal division include age, vigour, and pressure. The rate of division tends to be higher in young stems than in older ones (Bannan, 1950, 1960b; Srivastava, 1963b; Cumbie, 1967). Bannan (1950) found that, in vigorous young stems of *Chamaecyparis*, there were as many as three or four successive divisions of a fusiform initial in one growing season, as compared with one in older but still vigorous stems. Frequency of anticlinal division is not a direct function of the required increase in cambial circumference, because many cells are lost from the cambium. An average of 4·5 divisions sufficed to add one fusiform initial to the cambium in young material as compared with forty-three in old material. Bannan (loc. cit.) found that the frequency of anticlinal division is also affected by growth rate; there was an average of three or four successive divisions per year in young, fast-growing stems as compared with one every two years in slow-growing stems of the same age. The frequency of division in samples from a young, vigorous stem, a young, slow-growing stem and an old, vigorous stem is illustrated in Fig. 2.3. If the frequency of anticlinal division is expressed in terms of the thickness of wood produced it is usually constant. For example, in *Thuja* Bannan (1960a) found that in trees over 1·5 dm in diam. an initial divides on average two or three times per cm of xylem increment, regardless of tree size or rate of growth. Exceptions may occur in young trees (Bannan, 1960a) and in narrow growth rings (Bannan 1950, 1960a, 1962, 1963a, 1965, 1966a), where the relative frequency may be higher. Bannan (1957a) found that the frequency of anticlinal division in the cambium of fluted stems was higher in concave than in adjoining convex sectors. He considered that the radial pressure to which the concave arcs of cambium are subjected was a factor tending to accelerate the frequency of division.

It has long been recognized that loss of fusiform initials from the cambium is a common occurrence, but it was not until Bannan (1950) observed that as a result of over 1,100 anticlinal divisions in the cambium of *Chamaecyparis* there was a net gain of only 162 functional fusiform initials that the extent of such loss was realized. The amount of loss of fusiform initials from the cambium depends on the frequency of anticlinal division and the rate of elongation of initials, together with the rate of increase in circumference of the cambium. This last is dependent on the increase in circumference of the

woody cylinder, which itself is determined by the rate of production of
secondary xylem through the periclinal division of fusiform cambial initials
and xylem mother cells. As many fusiform initials are retained in the cambium as are required for increase in circumference; if the number of initials
exceeds this requirement the remainder are lost.

The daughter fusiform cambial cells formed by anticlinal division undergo
various types of transformation. Some elongate and expand to become functional fusiform initials; some decline and either lose their generative capacity,
eventually maturing into more or less abnormal xylem or phloem elements, or
develop into ray initials; and some become transversely subdivided, with the
resulting segments becoming lost through maturation or further reduced to
ray initials.

Failing fusiform initials may be termed *declining initials* and the radial files
of their derivatives *declining tiers* (Srivastava, 1963a). The process of decline
has been described by Bannan (1950, 1951a, 1953), Whalley (1950), Evert
(1961), Cumbie (1963), and Srivastava (1963a, 1963b), and is illustrated in
Fig. 2.4. A declining initial usually becomes reduced in all dimensions, and

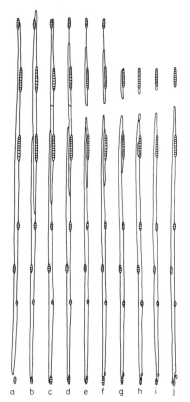

a   b   c   d   e   f   g   h   i   j

Fig. 2.4 Diagrams showing the sequence of cell
changes associated with the disappearance of a
fusiform initial, as seen in tangential sections; old
stem of *Chamaecyparis thyoides*. (Redrawn from
Bannan, 1950.)

the rate of periclinal division is often retarded. The initial shortens through successive periclinal divisions. It narrows tangentially, and the tangential walls may become thinner, while the bordered pits on the radial walls may be abnormally small and irregular in arrangement. Radial expansion tends to be greater in the centre than at the ends of the cell, resulting in a somewhat elongated spindle shape. Reduction in height is brought about by asymmetric periclinal division, in which the new tangential wall slices off to one side short of the cell tip, giving daughter cells of unequal lengths, with the shorter cell remaining as the functional initial. The declining initial may eventually cease to divide periclinally and be lost from the cambium, maturing into an abnormal xylem or phloem element. Alternatively, it may be further reduced in height and continue in the cambium as a ray initial, or it may divide transversely to give rise to several ray initials. Occasionally, fusiform initials are lost from the cambium abruptly, maturing into xylem or phloem elements without undergoing appreciable reduction in size, but more commonly the process is a gradual one.

Declining fusiform initials often undergo a series of transverse divisions, becoming segmented into a number of shorter units arranged in a vertical series. Segmentation may occur before or after reduction in height. The segments may be lost from the cambium by maturation, or they may become further reduced in size and develop into ray initials. There is some variation in the number of segments formed by the subdivision of a fusiform initial, but on the whole the number of segments appears to increase with increasing length of the initial. Bannan (1953) found that in certain conifers the relatively short fusiform initials of small branches or stems tended to subdivide only to the first or second order, while the larger initials of mature stems usually subdivided to the third order.

Whenever a fusiform initial is lost from the cambium the initials which converge from either side of it must form new wall contacts. The files of tracheids derived from these convergent initials develop matching pit pairs, and this development is often very rapid. Cambial initials and their immediate derivatives possess the ability to form new pit fields with altering cell contacts.

The causes of failure in a fusiform initial are not understood, but apparently unsatisfactory water relationships are involved. Declining initials seem unable to acquire or retain enough water for the cell expansion which normally accompanies periclinal division. Bannan regards the fusiform initials as competitive units in an overcrowded environment. In the competition for survival certain preferred types are selected, namely those with the greatest length and the most extensive contact with rays. There are indications that the ability to elongate is lost or decreased in short initials, while initials with the greatest ray contacts have an obvious advantage in the competition for water, food materials, and other substances necessary for growth.

The survival of fusiform initials may be under polar influence. Bannan (1968b), studying nearly 28,000 pseudotransverse divisions in twenty species of conifers, found that after about half these divisions one daughter cell survived as a functional fusiform initial while the other was lost. In some cambial sectors the upper of the two sister cells is more apt to persist, while in other sectors the reverse situation may hold. While there is much variation within a single tree, few species show approximately equal survival of upper and lower cells. Most species show greater survival among either upper or lower cells; where the upper cell usually persists, cell elongation is predominantly acropetal and where the lower one tends to survive basipetal elongation predominates. This correlation between survival after pseudotransverse division and the direction of major cell elongation indicates linkage to the same polar factor. Bannan suggests that certain substances concerned with viability and growth may be unequally distributed in fusiform initials, with a reversible tendency for a greater concentration towards one cell tip or the other.

The tendency for long fusiform initials to survive and short ones to decline has been noted by a number of authors (Bannan, 1951b, 1957c; Bannan and Bayly, 1956; Cheadle and Esau, 1964; Evert, 1961; Cumbie, 1963, 1969). Continued selection of the longest initials helps to maintain an adequate cell length in the cambium and its derivative tissues. Selection for length is apparent not only in the products of pseudotransverse division but also in cells cut off the sides of fusiform initials by lateral division. In *Thuja occidentalis* (Bannan, 1957b) very few cells produced in this manner survive as fusiform initials. A certain number are reduced but survive as ray initials, while the majority are lost through maturation. The mean length of the cells surviving as fusiform initials is much greater than that of the cells reduced to ray initials, which is in turn much greater than that of the cells which are lost from the cambium through maturation. In *Leitneria* and *Polygonum* a similar selection for length is apparent, with a tendency for shorter cells to be transformed into ray initials (Cumbie, 1967, 1969).

The extent of contact between newly formed fusiform initials and the cambial portions of the vascular rays is also important in survival (Bannan, 1951c, 1957c; Bannan and Bayly, 1956; Evert, 1961; Cheadle and Esau, 1964; Cumbie, 1967) (Fig. 2.5). In conifers studied by Bannan and Bayly surviving fusiform initials had contact with nearly 70% more ray cells than had those which failed. That the extent of ray contact exerts an independent effect on the survival of fusiform initials, and is not merely a result of greater cell length, is shown by a comparison on the basis of the ratio of ray cells to the length of the initial. Surviving initials had contact with 30% more ray cells per millimetre of length than had failing ones. Although very long cells usually survive and very short ones usually decline, in cells of intermediate length the extent of ray contact is a demonstrably important factor. Among such cells, those with the greatest ray contacts are the ones to survive. That the survivors are not necessarily the largest initials is shown by a comparison of surviving newly formed fusiform initials which were below the mean length for surviving cells with failing newly formed initials which were above the mean length for failing cells. The latter were slightly larger than the former, but had much less ray contact per millimetre of cell length.

The question arises whether length is significant in survival, apart from the additional ray contacts it allows. The importance of cell length is shown in the behaviour of very short newly formed initials. When a fusiform initial under-goes pseudotransverse division twice in rapid succession so that the mother cell is quartered, or when division is eccentric to give daughter cells of markedly uneven lengths, the short cells which result seldom survive as

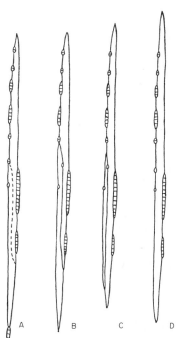

Fig. 2.5 Diagrams of cambial initials in tangential view showing contrasting behaviour of sister initials after origin in pseudotransverse division. A, pseudotransverse division; B, upper sister initial with extensive ray contacts, lower with sparse ray contacts; C, D, failure of lower sister initial and elongation of the upper initial into the 'space' vacated by the failing initial. (Redrawn from Bannan, 1957c.)

fusiform initials. If ray contacts are small they undergo rapid maturation or reduction to ray initials. Those with larger ray contacts may function for most of the growing season, but almost always fail before they can go through the usual cycle of enlargement and multiplication. Size, therefore, appears important in the survival of fusiform initials independently of the extent of ray contact.

The selection of fusiform initials on the basis of the extent of their contacts with rays plays an important part in the maintenance of cell pattern in the cambium. The processes of pseudotransverse division and elongation of the fusiform initials result in alterations in cell contacts, and ray deficiencies arise in some areas. The reduction of the fusiform cambial cells with more sparse ray contacts, often leading to the ultimate formation of one or more ray initials, helps to rectify these deficiencies and to maintain the required ratio of fusiform to ray cells.

The fact that newly formed fusiform initials follow certain patterns of behaviour as a general rule, but not without exception, points to the existence of a complex intercellular relationship in the cambium. Such deviations from the general trend as the loss of long fusiform initials with abundant ray contacts and the survival of shorter initials with poor ray contacts occur sporadically and apparently without reason.

The work of Bailey (1923) has indicated a line of development from structurally primitive dicotyledons with non-storeyed cambia and long fusiform initials which undergo pseudotransverse division and subsequent elongation to structurally advanced dicotyledons with storeyed cambia and short initials which undergo radial longitudinal division and do not elongate. Subsequent work has suggested further trends, although data are as yet too limited for definite conclusions to be drawn. Cumbie (1967) points out that there is a progressive series from the trees *Liriodendron tulipifera* and *Pyrus communis*, studied by Cheadle and Esau (1964) and Evert (1961) respectively, to the shrub *Leitneria floridana* and then to the herbaceous perennial *Hibiscus lasiocarpus*. Progressing along the series, length of initials decreases, elongation decreases, a smaller proportion of initials is lost and the frequency of lateral division increases, and in *Hibiscus* some divisions are radial longitudinal. All four species have non-storeyed cambia; loss of initials occurs, but is not common in species with short initials and storeyed cambia (Beijer, 1927; Butterfield, unpublished data). Conifers represent a stage which would precede the dicotyledonous series.

Cell elongation

In storeyed cambia increase in the cambial circumference is brought about by radial longitudinal division of the initials followed by their tangential expansion. In non-storeyed cambia the radial divisions of the fusiform initials are obliquely transverse, and the daughter cells subsequently elongate so that each may reach or exceed the length of the parent cell. Such elongation brings about an increase in the number of initials which intersect a given transverse plane and therefore increases the circumference of the cambium.

Detailed studies of the elongation of fusiform initials have been made in *Chamaecyparis* (Bannan and Whalley, 1950) and *Thuja* (Bannan, 1956). Perhaps the most striking feature of the elongation process is the high degree of variability that exists from cell to cell. This is due partly to differences in the inherent capacity of the cells to expand and partly to differences in the local environment. Variability in the growth patterns of the cells is superimposed upon, and tends to mask, a general trend which conforms to the familiar pattern of growth, with an initial phase of rapid elongation followed by a second phase where the growth rate gradually decreases as the cell lengthens.

The normal growth curve of fusiform initials may be greatly modified by local environmental conditions. Sometimes elongation proceeds in a continuous fashion, but often it shows sporadic acceleration and deceleration due to the influence of surrounding cells. When a neighbouring fusiform initial is lost from the cambium a growing initial usually elongates rapidly into the 'space' provided. Elongating tips often become stalled at rays, and a considerable delay may follow before the tip can thrust past the obstructing ray. In such cases the tip may become flattened, giving the end of the cell a truncate appearance. Occasionally brief flattening occurs when the tips of two elongating initials first meet.

In *Chamaecyparis* elongation following anticlinal division tends to be initially more rapid in the overlapping than the distal tips, but usually the distal tips catch up fairly rapidly, so that their combined elongation equals that in the overlap. The total amount of elongation may differ considerably, but the greater amount may be in either the region of overlap or the distal tips. There may be a substantial difference in the amount of elongation at opposite ends of the same initial; this is often due to conditions in the local environment, such as blocking rays and loss of neighbouring fusiform initials. Such differences are more common in *Thuja* (Bannan, 1956). Comparisons of elongation in the distal tips (i.e. the opposite ends of two sister initials) usually showed considerable differences.

Elongation is under polar influence in many conifer species (Bannan, 1968b). There is a tendency for elongation to be greater in one direction among groups of cambial cells, although the direction may be different in different sectors of the cambium. Most species show an overall predominance of either basipetal or acropetal elongation. Little information is available for dicotyledons, although Evert (1961) found no evidence of polar influence on elongating fusiform initials in *Pyrus communis*.

During the final phase in the growth cycle of fusiform initials, when elongation is slowing down, Bannan observed occasional recession of cell tips in *Thuja*. This sometimes becomes more common and pronounced with decline in the annual xylem production. The tip at the opposite end from the receding one may show no alteration or, rarely, shorten slightly. Usually it continues to elongate. Bannan points out that a cell which is receding at one end and elongating at the other is altering its position in the cambium. Such alteration may appear to be vertical displacement due to outside pressures, such as suggested by Priestley (1930) for differentiating hardwood fibres, but, in *Thuja* at least, this is not the case. The terminal shortening of fusiform initials is a gradual process (cf. reduction to ray initials) which may continue for two or three decades in slow-growing stems. This gradual recession is apparently achieved by the remoulding and slight shortening of the cell tip during the period of radial expansion between periclinal divisions.

The manner of elongation of cambial initials (and other types of cell) and their relationships with surrounding cells have been the subject of some controversy. Three principal theories have been advanced, namely, those of sliding, symplastic and intrusive growth.

In 1886 Krabbe suggested that, in meristems in general, each cell grew independently. Growth resulted in changes in the positions of cells relative to one another and the formation of new cell contacts. This Krabbe termed sliding or gliding growth (*gleitende Wachstum*); it was applied to the cambium by numerous workers, with the modification that elongation and the slipping of walls past one another occurred only towards the ends of the cells.

Priestley (1930a) felt that sliding growth was difficult to visualize and did not account satisfactorily for all the phenomena observed in the cambium and its derivatives. In particular, if walls slid past one another plasmodesmata would be broken and pits would either fail to develop or, if present, would not correspond on adjoining cell walls. Priestley applied his theory of symplastic growth to the elongation of cambial cells, suggesting that it fitted the known facts about the cambium better than the concept of sliding growth. Symplastic growth involves 'slow mutual adjustments of cell positions, with changes in cell size and shape, associated with a gradual adjustment of the walls as a common framework, without slip between cell and cell' (loc. cit., p. 121). Neighbouring primary cell walls grow at the same rate; the walls and the middle lamella grow as a common 'three-ply' membrane, with no alteration in previously formed cell contacts.

Symplastic growth is a gradual process, and therefore can be applicable only to a meristem in which changes in cell arrangement take place slowly. Until recently such a view of the cambium was in accord with the known facts and was generally accepted as correct. However, the work of Bannan and Whalley in 1950, together with later studies by Bannan and others on both conifers and dicotyledons, has shown quite clearly that earlier interpretations of the cambium as a gradually changing meristem were generally very far from the truth. The pattern of cells in the non-storeyed cambium is changing constantly, and often rapidly. The cambial initials divide anticlinally far more frequently than is necessary to keep pace with increase in cambial circumference; many of the fusiform initials so produced are lost from the cambium or converted to ray initials. The basic growth pattern of fusiform initials shows early rapid elongation, which later slows down, rather than the continuous slow extension envisaged by Priestley. Growth is not spread uniformly over the entire wall, but is localized towards the tips. Rates of cell elongation tend to be erratic and greatly influenced by the local environment. Individually, the initials behave like competitive units, the largest and most advantageously placed thriving, while smaller or less well-placed ones fail. Thus the theory of symplastic growth is clearly not applicable to the cambium, at least in the non-storeyed condition.

In 1939 Sinnott and Bloch, working with grasses, made direct observations of the apical root meristem. They found that different parts of the cell wall extended at different rates, one end of the cell being affected by a 'wave' of elongation before the other. Wall growth was sharply localized, being confined to the point of intrusion between other cells so that no slip occurred along previously formed contact faces. The cambium was not studied, but it was suggested that the fusiform initials elongated by growth at the tips only, and the term 'intrusive growth' was proposed.

While it seems to be established that fusiform initials elongate only at the tips, the exact amount of tip involved is not known. Growth of the cambium cannot be observed directly; the behaviour of a particular group of initials must be deduced from studies of its derivative cells. Since elongation occurs in the derivatives as well as the initials, conclusions about elongation in the initials are difficult to reach. Bannan and Whalley (1950), using serial tangential sections of the secondary xylem of *Chamaecyparis*, found no evidence to support the theory of symplastic growth in the cambium. On the contrary, the sequence of changes in the secondary xylem indicated that growth of each initial occurred independently of its neighbours. Growth in length appeared to be limited to the terminal parts of the initials, but the precise portion of tip involved could not be determined. Radial sections of the secondary xylem showed that the pits in the radial walls of the tracheids were in approximate radial alignment in the median portions of a file of cells, but were more irregular towards the ends. Such irregularity was probably due partly to elongation of the cambial initials and partly to growth during differentiation.

One of the problems which led Priestley to postulate symplastic growth in cambial cells was that of pit distribution. Along walls which move in relation to one another, plasmodesmata will be shorn and pit-fields displaced. Bannan and Whalley point out that the assumption that intruding cell tips are unable to form new pit-fields matching those on the stationary walls on either side is unjustified. Pits are nearly always more numerous towards the tips of the tracheids where elongation has occurred, than in the median portions where it has not, and they correspond with those on adjoining walls to form pit-pairs. Moreover, when a radial file of tracheids ceases due to the loss of a fusiform initial the adjacent walls of the converging tracheids do not lack matching pits, although the numbers of pits in the median parts of the convergent walls are sometimes reduced temporarily. It would thus seem that new pits may form not only in elongating walls but also in non-elongating ones, and that the cambial cell wall is a highly adaptable structure.

It may be concluded that the theory of symplastic growth is not applicable to non-storeyed cambia, in which the arrangement of cells is undergoing constant and rapid alteration. It may perhaps be applied to storeyed cambia, where cell arrangement is much more stable (Chapter 5). The distinction between the two remaining theories of sliding and intrusive growth is not altogether clear. It has never been suggested that entire cambial cells slide past one another; Krabbe (loc. cit.) considered that growth was mainly localized at the points where the initial penetrated between neighbouring cells. The exact extent of wall involved is still not known. If there is any real distinction it is perhaps that intrusive growth is restricted to the very tips of the cells and no slip occurs along previously formed contact faces. This latter is difficult to visualize, since, as Bannan (1956) points out, area growth of the wall involves the expansion of a structure already in existence. It would seem therefore that some slip must occur. Since the part of the wall involved is undergoing change and since the entire cambial wall is clearly very adaptable, this is quite feasible.

REFERENCES

BAILEY, I. W. (1919). Phenomena of cell division in the cambium of arborescent gymnosperms and their cytological significance. *Proc. Natl. Acad. Sci.* **5**, 283–5.
— (1920a). The formation of the cell plate in the cambium of higher plants. *Proc. Natl. Acad. Sci.* **6**, 197–200.
— (1920b). The cambium and its derivative tissues. II. Size variations of cambial initials in gymnosperms and angiosperms. *Am. J. Botany* **7**, 355–67.
— (1920c). The cambium and its derivative tissues. III. A reconnaissance of cytological phenomena in the cambium. *Am. J. Botany* **7**, 417–34.
— (1923). The cambium and its derivative tissues. IV. The increase in girth of the cambium. *Am. J. Botany* **10**, 499–509.
— (1930). The cambium and its derivative tissues. V. A reconnaissance of the vacuome of living cells. *Z. Zellforsch. Mikrosk. Anat.* **10**, 651–82.
BANNAN, M. W. (1950). Frequency of anticlinal divisions in fusiform cambial cells of *Chamaecyparis*. *Am. J. Botany* **37**, 511–17.
— (1951a). The reduction of fusiform cambial cells in *Chamaecyparis* and *Thuja*. *Can. J. Botany* **29**, 57–67.
— (1951b). The annual cycle of size changes in the fusiform cambial cells of *Chamaecyparis* and *Thuja*. *Can. J. Botany* **29**, 421–37.

BANNAN, M. W. (1953). Further observations on the reduction of fusiform cambial cells of *Thuja occidentalis* L. *Can. J. Botany* **31**, 63–74.

— (1955). The vascular cambium and radial growth in *Thuja occidentalis* L. *Can. J. Botany* **33**, 113–38.

— (1956). Some aspects of the elongation of fusiform cambial cells in *Thuja occidentalis* L. *Can. J. Botany* **34**, 175–96.

— (1957a). Girth increase in white cedar stems of irregular form. *Can. J. Botany* **35**, 425–34.

— (1957b). The relative frequency of the different types of anticlinal divisions in conifer cambium. *Can. J. Botany* **35**, 875–84.

— (1957c). The structure and growth of the cambium. *TAPPI* **40**, 220–5.

— (1960a). Cambial behaviour with reference to cell length and ring width in *Thuja occidentalis* L. *Can. J. Botany* **38**, 177–83.

— (1960b). Ontogenetic trends in conifer cambium with respect to frequency of anticlinal division and cell length. *Can. J. Botany* **38**, 795–802.

— (1962). Cambial behaviour with reference to cell length and ring width in *Pinus strobus* L. *Can. J. Botany* **40**, 1057–62.

— (1963a). Cambial behaviour with reference to cell length and ring width in *Picea*. *Can. J. Botany* **41**, 811–22.

— (1963b). Tracheid size and rate of anticlinal divisions in the cambium of *Cupressus*. *Can. J. Botany* **41**, 1187–97.

— (1964a). Tracheid size and anticlinal divisions in the cambium of *Pseudotsuga*. *Can. J. Botany* **42**, 603–31.

— (1964b). Tracheid size and anticlinal divisions in the cambium of lodgepole pine. *Can. J. Botany* **42**, 1105–18.

— (1965). The rate of elongation of fusiform initials in the cambium of Pinaceae. *Can. J. Botany* **43**, 429–35.

— (1966a). Cell length and rate of anticlinal division in the cambium of the sequoias. *Can. J. Botany* **44**, 209–18.

— (1966b). Spiral grain and anticlinal divisions in the cambium of conifers. *Can. J. Botany* **44**, 1515–38.

— (1968a). Anticlinal divisions and the organization of conifer cambium. *Botan. Gaz.* **129**, 107–13.

— (1968b). Polarity in the survival and elongation of fusiform initials in conifer cambium. *Can. J. Botany* **46**, 1005–8.

— and BAYLY, I. L. (1956). Cell size and survival in conifer cambium. *Can. J. Botany* **34**, 769–76.

— and WHALLEY, B. E. (1950). The elongation of fusiform cambial cells in *Chamaecyparis*. *Can. J. Res. C.* **28**, 341–55.

BARGHOORN, E. S., JR. (1940a). Origin and development of the uniseriate ray in the Coniferae. *Bull. Torrey Bot. Club* **67**, 303–28.

— (1940b). The ontogenetic development and phylogenetic specialization of rays in the xylem of dicotyledons. I. The primitive ray structure. *Am. J. Botany* **27**, 918–28.

BRAUN, H. J. (1955). Beiträge zur Entwicklungsgeschichte der Markstrahlen. *Bot. Stud., Jena* **4**, 73–131.

BUVAT, R. (1956). Variations saisonnières du chondriome dans le cambium de *Robinia pseudoacacia. C.R. Acad. Sci.* **243**, 1908–11.

CATESSON, A. M. (1961). Variations saisonnières du chondriome dans le cambium d'*Acer pseudoplatanus. C.R. Acad. Sci.* **252**, 2588–90.

— (1962). Modifications saisonnières des vacuoles et variations de la pression osmotique dans le cambium d'*Acer pseudoplatanus. C.R. Acad. Sci.* **254**, 3887–9.

— (1964). Origine, fonctionnement et variations cytologiques saisonnières du cambium de l'*Acer pseudoplatanus* L. (Acéracées). *Ann. Sci. nat. (Bot.). 12e ser.* **5**, 229–498.

CHEADLE, V. I. and ESAU, K. (1964). Secondary phloem of *Liriodendron tulipifera. Univ. Calif. Publ. Bot.* **36**, 143–252.

CÔTÉ, W. A. (ed.) (1965). *Cellular Ultrastructure of Woody Plants,* Syracuse University Press, New York.

CRONSHAW, J. and WARDROP, A. B. (1964). The organization of cytoplasm in differentiating xylem. *Australian J. Botany* **12**, 15–23.

CUMBIE, B. G. (1963). The vascular cambium and xylem development in *Hibiscus lasiocarpus. Am. J. Botany* **50**, 944–51.

— (1967). Developmental changes in the vascular cambium of *Leitneria floridana. Am. J. Botany* **54**, 414–24.

— (1969). Developmental changes in the vascular cambium of *Polygonum lapathifolium. Am. J. Botany* **56**, 139–46.

DODD, J. D. (1948). On the shapes of cells in the cambial zone of *Pinus sylvestris* L. *Am. J. Botany* **35**, 666–82.

EDGAR, E. (1961). *Fluctuations in Mitotic Index in the Shoot Apex of* Lonicera nitida, Univ. Canterbury Publs. no. 1. Christchurch, N.Z.

ESAU, K., CHEADLE, V. I. and GILL, R. H. (1966). Cytology of differentiating tracheary elements. I. Organelles and membrane systems. *Am. J. Botany* **53**, 756–64.

EVERT, R. F. (1961). Some aspects of cambial development in *Pyrus communis. Am. J. Botany* **48**, 479–88.

GREGORY, R. A. and WILSON, B. F. (1968). Cambial activity in white spruce: comparison between Alaska and New England. *Can. J. Botany* **46**, 733–4.

HARTIG, R. (1895). Ueber der Drehwuchs der Kiefer. *Sber. bayer. Akad. Wiss.* **25**, 199–217.

HEJNOWICZ, Z. (1961). Anticlinal divisions, intrusive growth and loss of fusiform initials in nonstoried cambium. *Acta Soc. Bot. Pol.* **30**, 729–58.

— (1964). Orientation of the partition in pseudotransverse division in cambia of some conifers. *Can. J. Botany* **42**, 1685–91.

— and BRANSKI, S. (1966). Quantitative analysis of cambium growth in *Thuja. Acta Soc. Bot. Pol.* **35**, 395–400.

HODGE, A. J. and WARDROP, A. B. (1950). An electron microscopic investigation of the cell wall organization of conifer tracheids and conifer cambium. *Australian J. Sci. Res. B* **3**, 265–9.

HOHL, H. R. (1960). Ueber die submikroskopische struktur normaler und hyperplastischer Gewebe von *Datura stramonium* L. I. Normalgewebe. *Ber. schweiz. Bot. Ges.* **70**, 395–439. (Summary in English.)

KERR, T. and BAILEY, I. W. (1934). The cambium and its derivative tissues. X. Structure, optical properties and chemical composition of the so-called middle lamella. *J. Arn. Arbor.* **15**, 327–49.

KLINKEN, J. (1914). Ueber das gleitende Wachstum der Initialen im Kambium der Koniferen und den Markstrahlenverlauf in ihrer secundären Rinde. *Bibl. Bot.* **19**, 1–37.

KRABBE, G. (1886). *Das gleitende Wachstum bei der Gewebebildung der Gefässpflanzen*, Bornträger, Berlin. Cited after Scott, 1888.

NÄGELI, C. (1864). *Dickenwachstum des Stengels und Anordnung der Gefässtränge bei den Sapindaceen*, München. Cited after Bailey, 1923.

NEEFF, F. (1920). Ueber die Umlagerung der Kambiumzellen beim Dickenwachstum der Dikotylen. *Z. Bot.* **12**, 225–52.

NEWMAN, I. V. (1956). Pattern in meristems of vascular plants. I. Cell partition in living apices and in the cambial zone in relation to the concepts of initial cells and apical cells. *Phytomorphology* **6**, 1–19.

PRESTON, R. D. and RIPLEY, G. W. (1954). An electron microscopic investigation of the walls of conifer cambium. *J. Exp. Botany* **5**, 410–13.

— and WARDROP, A. B. (1949). The sub-microscopic organization of the walls of conifer cambium. *Biochim. Biophys. Acta* **3**, 549.

PRIESTLEY, J. H. (1930a). Studies in the physiology of cambial activity. II. The concept of sliding growth. *New Phytol.* **29**, 96–140.

— (1930b). Studies in the physiology of cambial activity. III. The seasonal activity of the cambium. *New Phytol.* **29**, 316–55.

RAATZ, W. (1892). Die Stabbildungen im secundären Holzkörper der Bäume und die Initialentheorie. *Jarhb. wiss. Bot.* **23**, 567–636.

ROBARDS, A. W. and KIDWAI, P. (1969). A comparative study of the ultrastructure of resting and active cambium of *Salix fragilis* L. *Planta* **84**, 239–49.

SANIO, K. (1873). Anatomie der gemeinen Kiefer (*Pinus sylvestris* L.). *Jahrb. wiss. Bot.* **9**, 50–126.

SCOTT, D. H. (1888). Review of Krabbe. *Ann. Bot.* **2**, 127–36.

SINNOTT, E. W. and BLOCH, R. (1939). Changes in intercellular relationships during the growth and differentiation of living plant tissues. *Am. J. Botany* **26**, 625–34.

SRIVASTAVA, L. M. (1963a). Secondary phloem in the Pinaceae. *Univ. Calif. Publ. Bot.* **36**, 1–42.

— (1963b). Cambium and vascular derivatives of *Ginkgo biloba*. *J. Arn. Arbor.* **44**, 165–92.

— (1966). On the fine structure of the cambium of *Fraxinus americana* L. *J. Cell. Biol.* **31**, 79–93.

— and O'BRIEN, T. P. (1966). On the ultrastructure of the cambium and its derivatives. I. Cambium of *Pinus strobus* L. *Protoplasma* **61**, 257–76.

THIMANN, K. V. and KAUFMAN, D. (1958). Cytoplasmic streaming in the cambium of white pine. In THIMANN, K. V. (ed.), *The Physiology of Forest Trees*, pp. 479–92, Ronald Press, New York.

WARDROP, A. B. (1954). The mechanism of surface growth involved in the differentiation of fibres and tracheids. *Australian J. Botany* **2**, 165–75.

WHALLEY, B. E. (1950). Increase in girth of the cambium in *Thuja occidentalis* L. *Can. J. Res. C.* **28**, 331–40.

WILSON, B. F. (1964). A model for cell production by the cambium of conifers. In ZIMMERMAN, M. H. (ed.), *The Formation of Wood in Forest Trees*, pp. 19–36, Academic Press, New York.

— (1966). Mitotic activity in the cambial zone of *Pinus strobus*. *Am. J. Botany* **53**, 364–72.

— and HOWARD, R. A. (1968). A computer model for cambial activity. *Forest Sci.* **14**, 77–90.

ZIMMERMAN, M. H. (ed.) (1964). *The Formation of Wood in Forest Trees*, Academic Press, New York.

3
The origin and development of vascular rays

Almost all secondary vascular tissues contain horizontal strands of parenchyma known as vascular rays. These rays are essential for the functioning of
the vertical conducting elements of the wood, and a suitable ratio must be
maintained between the two in both amount and distribution. Thus, as the
cambial cylinder increases in diameter and new fusiform initials are formed,
additional ray initials must be produced. New groups of ray initials arise in
the cambium and enter a cycle of development in which they increase in size
and eventually split up into smaller units, which may again increase in size.
This tends to maintain a constant relationship between fusiform and ray
initials and ensure that ray tissue is spread evenly through the xylem.

The origin and development of vascular rays in both conifers and dicotyledons has been fully described by Barghoorn (1940a, 1940b, 1941a,
1941b). A more recent paper by Braun (1955) deals with ray development in
both groups. Other studies include those of Klinken (1914) on *Taxus baccata,* Bannan on *Chamaecyparis* (1950b, 1951) and *Thuja* (1941, 1951,
1953, 1956), Whalley (1950) on *Thuja,* Srivastava on the Pinaceae (1963a)
and *Ginkgo biloba* (1963b), Beijer (1927) on *Aeschynomene* and *Alstonia,*
Chattaway (1933, 1938) on the Sterculiaceae, Evert (1959, 1961) on *Pyrus
communis,* Cumbie on *Hibiscus lasiocarpus* (1963), *Leitneria floridana*
(1967) and *Polygonum lapathifolium* (1969), and Cheadle and Esau (1964)
on *Liriodendron tulipifera.*

The origin of rays

Vascular rays may be divided into two categories according to their place of
origin. Certain cells of the primary body, the primordial ray initials, give rise
to ray initials which produce the primary rays. Primary rays may thus be

defined as rays occurring in the secondary body but originating in the primary tissues. Secondary rays, on the other hand, originate during the development of the secondary body. The ray initials which produce them are formed from the fusiform initials of the cambium.

Primary rays arise from both the fascicular and the interfascicular regions of the stele of the primary body. In conifers the primordial ray initials are parenchymatous and more or less regularly rectangular, with their long axes in the vertical plane. In the interfascicular regions they comprise cells which have not elongated to any marked degree after their production from the apex. In the fascicular regions, however, they are products of the segmentation of extensively elongated cells of the procambial strands. The vascular rays of conifers are all uniseriate, with similar primary rays being formed from the fascicular and interfascicular regions (Fig. 3.1). The primary rays are charac-

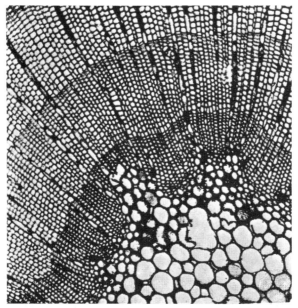

Fig. 3.1 Keteleeria davidiana (Bertr.) Beissn. Transverse section of the stem showing relation of primary ray origins to the fascicular and interfascicular segments of the primary body. (From Barghoorn, 1940a.)

teristically very low, and where there is a long vertical series of primordial ray initials certain of these may fail to develop, leading to the formation of several low rays rather than a single high one. A second factor contributing to the formation of low rays is the shortening of the primordial ray cells as they develop into initials. This brings about the separation of vertically contiguous cells and the development of several low rays (Thomson, 1914; Bannan, 1934; Barghoorn, 1940a). The first cells of the primary rays are often very irregular in shape and frequently have long projections (Fig. 3.2), features characteristic of the extracambial products of asymmetric periclinal division.

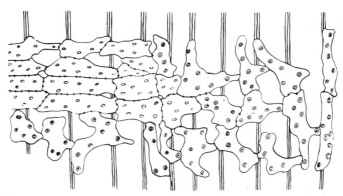

Fig. 3.2 Cedrus libanitica Trew. Radial longitudinal section showing origin of a ray in the root and transition from vertically to radially elongated ray cells. Note amoeboid appearance of ray cells. (From Barghoorn, 1940a.)

In the first-formed secondary vascular tissues of roots the transition in the shape of the ray initials from vertically elongate to radially elongate is slower than in stems, allowing time for transverse anticlinal divisions to occur in the initials. Fewer primordial ray initials fail to develop in roots, and because of these two factors vertical contiguity tends to be maintained between the individual cells of the ray (Fig. 3.2), as in dicotyledons (Fig. 3.4).

Structurally primitive dicotyledons possess both uniseriate and multiseriate rays. The uniseriate ray is high and composed of large, vertically elongated cells, while the multiseriate one has a main body made up of more or less isodiametric cells and long wings of cells similar to those of the uniseriate ray (Fig. 3.14A). The origin of primary rays in such forms is closely connected with the segmentation of the stele of the primary body into fascicular and interfascicular regions (Fig. 3.3). The very high, narrow uniseriates arise through the septation of very long, radially and tangentially narrow procambial cells of the vascular bundles to form long strands of ray initials. The primary multiseriate rays arise through the formation of ray initials from the cells of the interfascicular region, which have not undergone much elongation during their differentiation from the apical meristem and tend to produce initials more nearly isodiametric than those produced from the procambium of the vascular bundles. In contrast to the situation in conifer stems, the primary uniseriate rays are seldom reduced in height as a result of radial

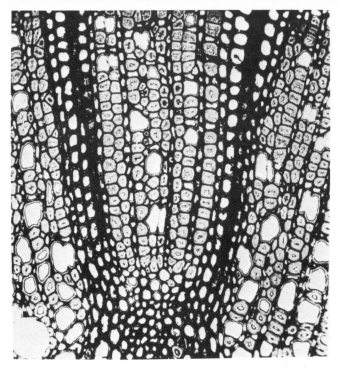

Fig. 3.3 Villarsia mucronata Ruiz & Pav. Transverse section showing two primary multiseriate rays extending from interfascicular segments of the stele. Three primary uniseriate rays
extend from the fascicular segment. (From Barghoorn, 1940b.)

extension of the initials. If a change occurs in the orientation of the long axis
from the vertical to the radial plane it is preceded by transverse anticlinal
division, and the shorter initials maintain their normal cell contacts during
radial expansion (Fig. 3.4).

With the production of secondary vascular tissue from the cambium, the
stem increases in circumference and the primary rays grow farther and
farther apart. While this is taking place new rays are formed from the
cambium in the region between the primary rays. These are the secondary
rays, and as the stem continues to grow in circumference their formation
contributes to the maintenance of a more or less constant ratio of vertical to
horizontal tissues in the vascular cylinder.

Secondary rays arise from fusiform initials. Ray initials are formed in
various ways: a single cell may be cut off the side or the end of a fusiform
initial; a declining fusiform initial may be reduced to a single ray initial or all
or part of a fusiform initial may be reduced to a single ray initial, or all or
part of a fusiform initial may be segmented by transverse divisions to form a

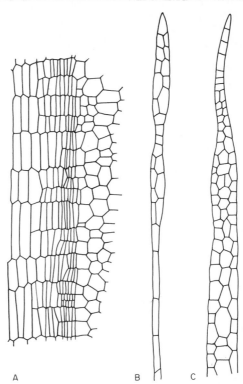

Fig. 3.4 Vascular ray of *Hoheria angustifolia* Raoul. A, R.L.S. of ray shortly after secondary thickening has begun. The cambium runs down the centre of the figure: its initials have been divided by repeated horizontal divisions. Few derivatives have been added to the outside. The isodiametric cells to the right represent part of a medullary ray in the primary phloem. B, T.L.S. of ray initials at an earlier stage than A. C, T.L.S. of ray initials at the same stage of development as A.

tier of potential ray initials (Fig. 3.5). The origin of ray initials from radial plate initials has been described by Barghoorn (1940a). Radial plates are radially arranged sheets of parenchyma characteristic of the secondary phloem of some conifers. Their initials have their long axes in the vertical plane; they produce derivatives mainly in the phloem and are usually lost after only a short period of activity. When they border on a group of ray initials they may, however, continue to function indefinitely, producing a border of erect cells in the ray. In a group of radial plate initials not associated with a ray one (or more) of the central cells may not be lost, but may continue to function as a ray initial. Radial plates are, in fact, a passing phase in the reduction of fusiform initials and the origin of rays (Chrysler, 1913; Bannan, 1953; Srivastava, 1963a). The 'radial plate initials' are slowly

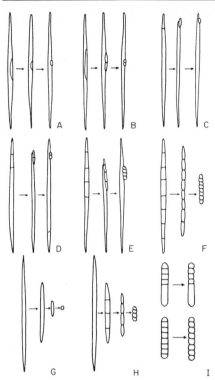

Fig. 3.5 Diagrammatic representation of the origin of rays from fusiform initials. A, lateral division with the formation of a single ray initial; B, lateral division with the formation of several ray initials; C, division at the end with the formation of one ray initial; D, division at the end with the formation of several ray initials; E, subdivision of a half-initial to form a vertical tier of ray initials; F, subdivision of an entire fusiform initial; G, reduction in height of a declining fusiform initial with the formation of a single ray initial; H, reduction in height followed by subdivision of a declining fusiform initial to form a vertical tier of ray initials; I, subdivision of one half of a short fusiform initial to form a tier of ray initials without reduction in the height of the segments; J, subdivision of an entire, short fusiform initial to form a tier of ray initials without reduction in height of the segments. (In C, D, and E the position of the ray appears to alter due to the elongation of the fusiform initial.)

declining fusiform initials and a radial plate is a declining tier in the phloem. Thus the origin of ray initials from radial plate initials is nothing more than the later part of the process whereby ray initials are derived from declining, segmented fusiform initials.

Although the formation of ray initials may be conveniently described as occurring in four ways (division off the side, division off the end, decline, and segmentation), these are not entirely distinct from one another. In the formation of ray initials at the side of a fusiform initial the division typically occurs about halfway along the cell, with the cell plate twice intersecting the same wall. The smaller daughter cell may be reduced in height before its conversion to one or more ray initials. All degrees of transition occur between this type and the formation of ray initials from a cell cut off the end of a fusiform initial, in which the division may occur at varying distances from the end of the fusiform initial with the cell plate twice intersecting the same side-wall or, alternatively, may take place at the end of the fusiform initial, cutting off the entire tip and so decreasing its length (Barghoorn, 1940a).

The daughter cells formed by the pseudotransverse division of a fusiform initial may elongate to become functional fusiform initials or, alternatively, they may decline and/or be transformed into ray initials. A declining cell may decrease in size and either mature into a xylem or phloem element or develop into a single ray initial. Alternatively, reduction in size may be accompanied by one or more transverse divisions. The segments of the sub-divided cell may continue to decline and eventually disappear from the cambium, or they may develop into ray initials. In most cases some of the segments are lost from the cambium while others develop into ray initials, but the proportion of surviving cells is highly variable, and sometimes all may be lost or all may survive. A fusiform initial which becomes segmented by a series of transverse divisions may undergo reduction in height either before or after segmentation. In cambia with short fusiform initials there may be no reduction in height and no loss of segments from the cambium.

Fusiform initials sometimes undergo eccentric pseudotransverse division (Bannan, 1950b, 1951; Whalley, 1950; Srivastava, 1963b; Cumbie, 1963, 1967), in which the cell plate is formed somewhat away from the centre of the cell. The smaller daughter cell may decline and/or segment to produce ray initials.

In the origin of rays through segmentation of fusiform initials the number of fusiform cells involved in the formation of a ray varies. After pseudotransverse division either one or both daughter cells may subdivide to produce ray initials. Occasionally several vertically and tangentially contiguous fusiform cells may subdivide simultaneously in dicotyledonous cambia, leading to the formation of a new multiseriate ray. A major factor affecting the formation of ray initials from segmented fusiform initials is the ray contacts of the latter. Ray initials tend to arise through the segmentation of fusiform initials which have poor ray contacts, and in this way a constant relationship between fusiform and ray elements is maintained.

Cells which are formed by the subdivision of fusiform initials usually have their long axes in the vertical plane. Those destined to become ray initials usually become reduced in height. Bannan (1951) records that in *Chamaecyparis* and *Thuja* the height of the individual segments is at first generally 150–500 μm and becomes reduced to the typical ray height of approximately 20 μm, with an alteration in the orientation of the long axis from the vertical to the radial plane. Similar adjustments may take place in cells cut off the ends or sides of fusiform initials. Reduction in height is brought about by further transverse anticlinal division and/or asymmetric periclinal division (Bannan, 1951, 1953, 1956). In the early stages of segmentation the rate of periclinal division may be retarded, and this is accompanied by a delay in volume increase. The radial expansion of the cell is often less at the upper and lower ends than at the middle, with the result that the cells become tapered at the ends and the tips pull apart. When periclinal division occurs, the new walls intersect the older tangential walls short of the cell tips, resulting in daughter cells of unequal lengths. The orientation of the cells is such that the shorter cell remains in the cambium, the larger one going into the xylem or phloem. A succession of such asymmetric divisions causes rapid

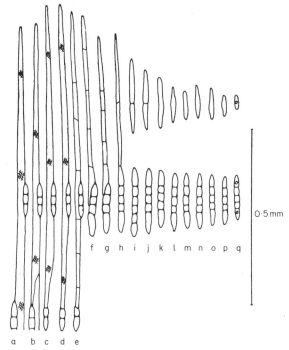

Fig. 3.6 Pinus pinea L., tangential sections showing successive phloic derivatives of a declining fusiform initial that eventually gave rise to three ray initials (shown with nuclei at *q*). The transverse wall, just above the three-celled ray, in the derivative at *e*, is interpreted as having resulted from a transverse division in the cambial initial; other transverse or oblique walls are results of division in the phloic initials. (Sieve cells with cross-hatched areas, derivatives of the declining fusiform initial and new ray initials without nuclei, cells in established rays without nuclei.) (From Srivastava, 1963a.)

reduction in height, and by this means a cell may be reduced to a small fraction of its original height in a very short period of time (Figs. 3.6, 3.7).

The relative frequency of the various types of ray origin in conifers is not altogether clear. Numerous workers have been of the opinion that most rays originate through divisions off the ends or sides of fusiform initials (Schmidt,

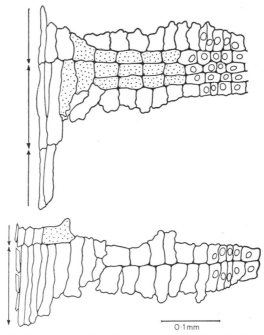

Fig. 3.7 Pinus pinea, radial views showing the formation of rays after the decline of fusiform initials. Arrows indicate segments of declining fusiform initials. (Albuminous cells without nuclei, cells with starch stippled, cells in the cambial zone with nuclei.) (From Srivastava, 1963a.)

1889; Klinken, 1914; Barghoorn, 1940a; Braun, 1955), although Barghoorn found origin through segmentation of fusiform initials to be quite frequent in young, rapidly growing stems. Bannan (1934, 1950b, 1951, 1953), Whalley (1950), and Srivastava (1963a), on the other hand, found that most rays originate through the decline and/or segmentation of fusiform initials. Bannan (1957) states that divisions off the sides of fusiform initials are infrequent in conifer cambium; this is supported by the observations of Whalley (1950) and Srivastava (1963a). Barghoorn (1940a) considers that rays frequently originate from a very small cell cut off the end of a fusiform initial. Bannan (1950b, 1951) has not observed this type of origin but found that rays occasionally arise when a considerable portion of one end of a fusiform initial is cut off. Dodd (1948), in an investigation on the shape of fusiform initials, found that some had truncate ends. He concluded that these had resulted from the tips being cut off by transverse walls. This would appear to support Barghoorn's views on ray origin, but Bannan and Whalley (1950) point out that such truncate ends could arise from contact between the elongating tips of fusiform initials and obstructing rays of earlier origin. Bannan (1951) noted a frequent association of new rays with tracheid tips in the wood of *Chamaecyparis* and *Thuja*; however, the ray initials did not originate from the tips of the fusiform initials which produced the adjoining tracheid rows, but were of earlier origin.

The high rate of production of new fusiform cells in the cambium, and the consequent failure of many of these to develop into fusiform initials, was not appreciated until the publication of Bannan's and Whalley's papers on *Chamaecyparis* and *Thuja* in 1950. Bannan (1950b, 1951) points out the importance of this phenomenon in considering the problem of ray origin. Most newly formed fusiform cambial cells which fail to develop into fusiform initials undergo subdivision to some degree, so that the number of potential ray initials present in the cambium as segments of subdivided fusiform cells is very great, and far in excess of the number required.

In some dicotyledons new uniseriate rays one or two cells high commonly arise by divisions off the ends or sides of fusiform initials. In divisions off the end a considerable portion of the tip is cut off in the origin of even a one-celled ray (Barghoorn, 1940b), and all stages exist between this and the origin of high rays by the segmentation of fusiform initials, so that the difference between origin by division off the end and origin by segmentation is one of degree rather than kind. Declining fusiform initials which are undergoing reduction in height may produce one or, by segmentation, several ray initials, or whole or half-length fusiform initials may be converted into ray initials by segmentation without reduction in height.

There appears to be a trend in the origin of rays through segmentation of fusiform initials, which is related to the already mentioned (p. 30) tendency towards decreasing loss of fusiform initials with increasing structural specialization and shortening of initials. In conifers and in the primitive dicotyledon *Liriodendron tulipifera* (Cheadle and Esau, 1964) the segments of the subdivided fusiform initials undergo reduction in height, and usually some of the segments are lost. In *Pyrus communis* (Evert, 1961) declining fusiform initials undergo height reduction and some products of segmentation may be lost. In *Leitneria floridana* (Cumbie, 1967) long initials may decrease in height and segment with the loss of some segments, but relatively short initials undergo segmentation without any appreciable reduction in height, and loss of segments is uncommon. In *Hibiscus lasiocarpus* and *Polygonum lapathifolium* (Cumbie, 1963, 1969) there is little or no reduction in height and usually no loss of segments, and in *Aeschynomene*, which has short initials and a storeyed cambium, there is no reduction in height of fusiform initials and apparently no loss of segments (Beijer, 1927; Butterfield, unpublished data).

Ray development in conifers

The constantly changing structure of the cambium is reflected in the changes which take place in the rays. In conifers these are relatively simple, with rays tending to develop towards a constant height through the increase in height of low rays and the decrease in height of high ones.

Increase in the height of rays may be brought about by increase in the size of the initials, by their transverse anticlinal division, or by ray fusion. Expansion of initials does not in itself bring about any very marked increase in ray height. Transverse anticlinal division accompanied by cell expansion is an important factor, and occurs frequently in isolated initials shortly after their origin. In rays more than three cells high such divisions are restricted to the marginal initials and become less frequent as the ray increases in height. Most conifer wood has very high rays, yet both primary and secondary rays are very low at the time of their origin. The effect of transverse anticlinal division is not great enough to bring about the transformation, which is effected largely by the fusion of rays. Rays frequently become fused through the loss of intervening fusiform initials from the cambium (Fig. 3.8B). If the two rays overlap, their fusion forms a partially biseriate ray which rapidly becomes uniseriate through the loss of ray initials. Fusion may be brought about by transverse anticlinal division, followed by cell expansion in the marginal initials of two vertically adjacent rays. It may result from cell expansion alone, but in this case is often temporary. Rays may also increase in height by the addition of segments from declining fusiform initials (Figs. 3.6, 3.8A).

Decrease in the height of rays is brought about by the splitting of rays and the loss of ray initials (Fig. 3.8C). Splitting occurs when an elongating fusiform initial intrudes between the initials of the ray. Barghoorn (1940a) found that loss of initials occurs in all parts of the ray, but is rare except in genera where reduction of ray tissue is extensive. He considered that the process of splitting is dependent on chance and that, since the probability of its happening therefore increases with the height of the ray, it tends to preserve a constant ray height. These observations are not in accord with those of Evert (1961) on the dicotyledon *Pyrus communis* and Srivastava (1963a) on the Pinaceae. Evert found that even when splitting appears to result from the intrusive growth of fusiform initials, this is invariably preceded by the loss of ray initials from the region where splitting occurs. Srivastava found only one instance in which intrusion was not preceded by loss of ray initials; in this case the ray initials decreased in height and separated from each other before the fusiform initial intruded between them.

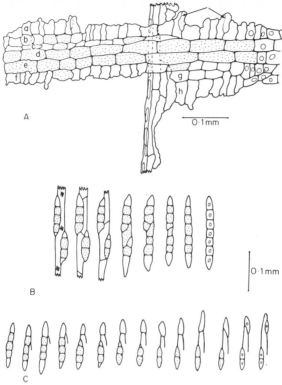

Fig. 3.8 Pinus pinea, sections of rays. A, radial section showing the addition of new ray initials at ray margins. Arrows indicate results of periclinal divisions in the phloic initials. Rows of ray cells are lettered. B, tangential sections showing the fusion of two rays after the loss of a segment of a declining fusiform initial between them. C, tangential sections showing the break-up of a ray by the loss of ray initials and the intrusion of a fusiform initial into the vacated space. (Sieve cells with cross-hatched areas, crystal cells with crystals, albuminous cells without nuclei, starch-containing cells stippled, cells in the cambial zone with nuclei.) (From Srivastava, 1963a.)

In contrast to the situation in many dicotyledons, the elongation of ray initials to fusiform size is not a factor in the splitting of normal conifer rays. Such transformations do not, except under unusual circumstances, occur in conifer cambium (Bannan and Bayly, 1956). Boureau (1956) records that they may occur in the elimination of rays, when all the initials of a ray elongate to fusiform size. Bannan (1941, 1942, 1944, 1950a) has described abnormal rays which occur sporadically in the cambium of some genera of the Cupressaceae (Fig. 3.9). Such rays begin at, or very close to, the pith, and may contain both parenchymatous and tracheary cells of varying lengths. They may become very wide and be dissected into smaller units by the

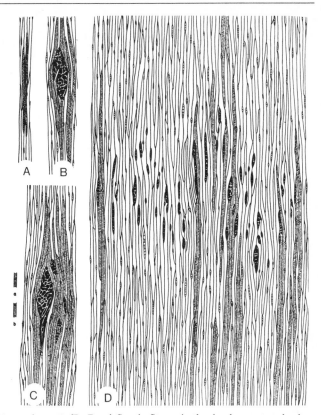

Fig. 3.9 Chamaecyparis nootkatensis (D. Don.) Spach. Stages in the development and subsequent dissection of an abnormal ray in a branch. The cross-lined tracheids belong to two lineal series beginning on either side of the ray in A. (From Bannan, 1950a.)

intrusion of fusiform initials or occasionally by the conversion of central ray initials to fusiform initials.

Ray development in structurally primitive dicotyledons

The development of rays is more complex in structurally primitive dicotyledons than in conifers, because multiseriate as well as uniseriate rays are present. The primary rays are typically very high at their origin and decrease in height during early secondary growth. This contrasts with the situation in conifers, where originally very low rays quickly increase in height. The primary multiseriate rays are reduced in width as they are dissected into smaller units, while the primary uniseriate rays broaden into multiseriates. Secondary rays continually arise in the cambium, usually as uniseriates; they increase in height and width and are eventually broken up into smaller multiseriates. Several types of cellular change are involved in ray development.

New ray initials arise through the division of fusiform or other ray initials. Both fusiform and ray initials increase in size. Ray initials are transformed into fusiform initials. Both ray and fusiform initials are lost from the cambium. Throughout such diverse and opposing activities, a pattern of ray and fusiform tissue characteristic of the species is maintained, indicating the plastic but highly co-ordinated nature of the cambium.

The rays of structurally primitive dicotyledons may increase in height and width by increase in the size of the ray initials, by division of the initials and subsequent expansion of the daughter cells, by the addition of fusiform initials to the ray, or by the fusion of rays. In contrast to those of conifers, the ray initials of dicotyledons may divide anticlinally, even if situated centrally in a ray, and the direction of division may be either longitudinal or transverse, leading to increase in width as well as height. An abrupt increase in the

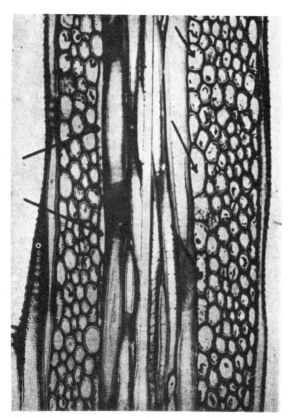

Fig. 3.10 Ilex opaca Ait. Tangential longitudinal section of xylem showing 'sheath cells' at the sides of the multiseriate rays, as indicated by arrows. These cells are formed by septation of fusiform initials added to the sides of the ray. (From Barghoorn, 1940b.)

dimensions of a ray, either vertically or tangentially, may occur by the addition of one or several fusiform initials to the ray. This process is similar to the origin of a ray from a subdivided fusiform initial, but occurs immediately next to an existing ray. The fusiform initial becomes segmented into a vertical series of small units which assume the character of ray initials. Some dicotyledons, such as certain members of the Sterculiaceae (Chattaway, 1933, 1938), possess a sheath of elongated cells around their rays (Fig. 3.10). This sheath is formed by the incorporation of subdivided fusiform initials on to the rays; the sheath initials remain elongated as long as they remain on the outside of the ray, but if the apposition of further subdivided fusiform initials places them in a central position they assume the character of ray initials. The characteristic wings of primitive multiseriate rays are formed in a similar manner. The fusion of either vertically or laterally approximated rays is brought about in a number of ways. Marginal ray initials may intrude between fusiform initials or grow into the space caused by their loss, and in doing this the ray initials may merely increase in size or they may also divide (Fig. 3.11). If all the fusiform initials lying between two neighbouring rays become incorporated into one or other of these rays this, too, may result in fusion.

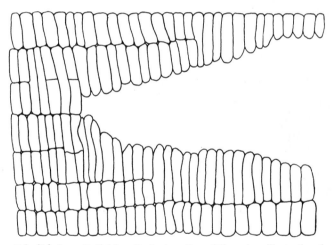

Fig. 3.11 Casearia nitida (L.) Jacq. Radial longitudinal section of the xylem illustrating the fusion of two rays by the vertical elongation of the marginal initials and their derivatives. Cambium to the left. (From Barghoorn, 1940b.)

Rays may decrease in size by the loss of ray initials or the splitting of rays. Splitting may be brought about by the intrusion of elongating fusiform initials between the initials of the ray (Fig. 3.12) or by the transformation of ray initials into fusiform initials (Fig. 3.13). The intrusion of fusiform initials may

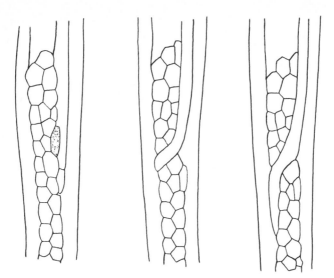

Fig. 3.12 Trochodendron aralioides Sieb. & Zucc. Serial tangential sections of the xylem showing the splitting of a ray by the apical elongation of a fusiform initial. The ray initial indicated by shading in the first figure was lost from the cambium, facilitating the entrance of the fusiform initial into the ray. (From Barghoorn, 1940b.)

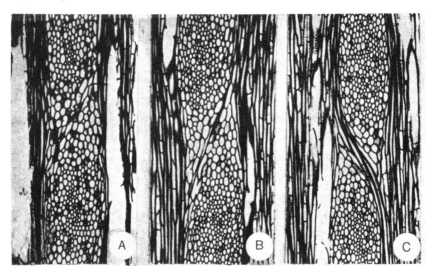

Fig. 3.13 Melicytus ramiflorus Forst. Series of three representative stages in the splitting of a ray by the transformation of ray initials to fusiform initials. (From Barghoorn, 1940b.)

reduce the width as well as the height of a multiseriate ray; its importance as a factor in ray dissection depends on the amount and rate of elongation of fusiform initials, which is subject to wide variation. Evert (1961) found that, in *Pyrus communis*, ray splitting by intruding fusiform initials is always preceded by loss of ray initials from the region of intrusion. *Liriodendron* is apparently similar to *Pyrus* in this respect (Cheadle and Esau, 1964). The conversion of ray initials to fusiform initials is common in dicotyledons, occurring most frequently in the central regions of the multiseriate rays. Generally, several adjacent ray initials elongate during successive periclinal divisions and gradually assume the character of fusiform initials. The long axes of the cells tend to be oriented parallel to one another and at an acute angle to the long axis of the ray, resulting in a dissection of the ray which reduces it in both height and width. The elongation of ray initials to fusiform size does not occur in all dicotyledons, and was not observed by Evert and Cheadle and Esau in *Pyrus* and *Liriodendron* respectively.

Modification of rays in dicotyledons

The dicotyledons show a considerable diversity of ray structure, with various lines of specialization from a primitive type (Kribs, 1935). Primitively, the uniseriate rays are high and numerous and made up of large, vertically elongated cells, while the multiseriate rays are high and wide, with a central body of vertically and tangentially short cells and uniseriate wings of cells similar to those of the uniseriate rays. From this primitive type modified forms have developed through reduction in the size of rays, a trend towards uniformity in the shape and size of the ray initials, the loss of uniseriate wings from the multiseriate rays and reduction in number or elimination of one or both types of ray (Fig. 3.14).

Reduction in the size of rays is brought about by the increased activity of the processes already described as decreasing the size of rays, together with a decrease in the activity of processes which bring about the enlargement of rays. Ray initials tend to become uniformly more or less isodiametric through the rapid vertical shortening of all cells which develop into ray initials. The uniseriate wings characteristic of primitive multiseriate rays tend to be absent from more advanced forms. This is partly because the fusiform initials tend to be shorter, and therefore give rise to wings less frequently when added to the sides of rays, and partly because when uniseriate wings are formed they become multiseriate very rapidly as a result of longitudinal anticlinal divisions in the initials.

Reduction in the number of uniseriate rays is brought about by a rapid increase in width immediately after their formation or by the suppression of their formation. In species in which uniseriate rays are completely absent secondary multiseriates may arise as such in the cambium. More commonly,

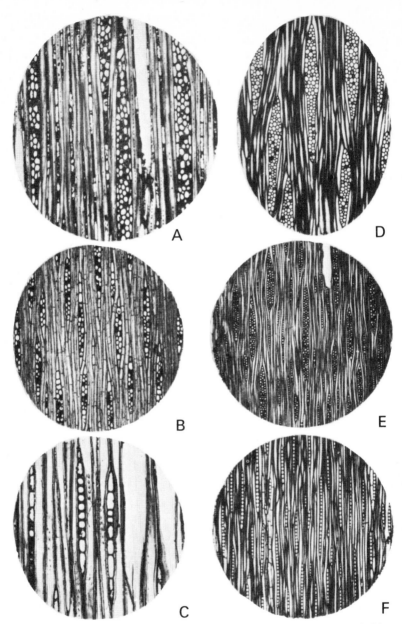

Fig. 3.14 Tangential sections showing types of wood ray in dicotyledons. A, primitive structure, *Taonabo peduncularis* (DC.) Pittier; B, reduction in ray size, *Dracontomelon dao* Merr.
& Rolfe; C, elimination of multiseriate rays, *Schima confertiflora* Merr; D, elimination of
uniseriate rays, loss of wings, rays homocellular, *Crataeva tapia* L.; E, homocellular rays, *Acer
mandshuricum* Maxim.; F, elimination of multiseriate rays, rays homocellular, *Castanea
pumila* (L.) Miller. (From Kribs, 1934.)

secondary rays are absent, and the even distribution of ray tissue is brought about by a repeated cycle of dissection and increase in size of the primary rays. The distance between the smaller units formed by dissection of the ray increases as the intervening fusiform initials multiply by anticlinal divisions and the cambium increases in circumference (as in Fig. 3.9).

Reduction in the number of multiseriate rays is brought about by the suppression of the processes which increase the width of rays. In species in which multiseriate rays are absent the primary rays, including those found in the interfascicular regions, are always uniseriate at their origin, as are the secondary rays. The ray initials do not undergo longitudinal anticlinal division.

A type of ray known as the aggregate ray occurs occasionally throughout the dicotyledons. Groups of uniseriate rays, or uniseriates and small multiseriates, occur together, forming the general outline of a very large multiseriate ray. The cambial initials of such a ray may comprise all stages of transition between typical ray initials and fusiform initials. The rays of the aggregation do not diverge with increase in cambial girth, because anticlinal divisions are suppressed in the fusiform cells included in them.

Ray tissue is almost or entirely absent from some dicotyledonous woods. This is correlated with reduced cambial activity and a tendency towards the herbaceous habit of growth. It occurs in a wide range of shrubs and semishrubs and is often associated with other cambial anomalies (Chapter 7). Rays are often suppressed, although rarely entirely absent, in plants which are dwarfed or otherwise modified for xerophytic environments. The elimination of rays may be brought about by the suppression of ray initial formation or by the transformation of ray initials into fusiform initials. If this tendency for elongation affects all the initials of a ray it loses its morphological identity, becoming similar to the surrounding fusiform tissue.

REFERENCES

BANNAN, M. W. (1934). Origin and cellular character of xylem rays in gymnosperms. *Bot. Gaz.* **96**, 260–81.

— (1941). Vascular rays and adventitious root formation in *Thuja occidentalis* L. *Am. J. Botany* **28**, 457–63.

— (1942). Wood structure of the native Ontario species of *Juniperus. Am. J. Botany* **29**, 245–52.

— (1944). Wood structure of *Libocedrus decurrens. Am. J. Botany* **31**, 346–51.

— (1950a). Abnormal xylem rays in *Chamaecyparis. Am. J. Botany* **37**, 232–6.

— (1950b). The frequency of anticlinal divisions in fusiform cambial cells of *Chamaecyparis. Am. J. Botany* **37**, 511–19.

BANNAN, M. W. (1951). The reduction of fusiform cambial cells in *Chamaecyparis* and *Thuja. Can. J. Botany* **29,** 57–67.

— (1953). Further observations on the reduction of fusiform cambial cells in *Thuja occidentalis* L. *Can. J. Botany* **31,** 63–74.

— (1956). Some aspects of the elongation of fusiform cambial cells in *Thuja occidentalis* L. *Can. J. Botany* **34,** 175–96.

— (1957). The relative frequency of the different types of anticlinal divisions in conifer cambium. *Can. J. Botany* **35,** 875–84.

— and BAYLY, I. L. (1956). Cell size and survival in conifer cambium. *Can. J. Botany* **34,** 769–76.

— and WHALLEY, B. E. (1950). The elongation of fusiform cambial cells in *Chamaecyparis. Can. J. Res. C.* **28,** 341–55.

BARGHOORN, E. S., JR. (1940a). Origin and development of the uniseriate ray in the Coniferae. *Bull. Torrey Bot. Club* **67,** 303–28.

— (1940b). The ontogenetic development and phylogenetic specialization of rays in the xylem of dicotyledons. I. The primitive ray structure. *Am. J. Botany* **27,** 918–28.

— (1941a). The ontogenetic development and phylogenetic specialization of rays in the xylem of dicotyledons. II. Modification of the multiseriate and uniseriate rays. *Am. J. Botany* **28,** 273–82.

— (1941b). The ontogenetic development and phylogenetic specialization of rays in the xylem of dicotyledons. III. The elimination of rays. *Bull. Torrey Bot. Club* **68,** 317–25.

BEIJER, J. J. (1927). Die Vermehrung der radialen Reihen im Cambium. *Rec. Trav. Bot. neerl.* **24,** 631–786.

BOUREAU, E. (1956). *Anatomie Végétale* **1,** U.P. de France, Paris.

BRAUN, H. J. (1955). Beiträge zur Entwicklungsgeschichte der Markstrahlen. *Bot. Stud., Jena* **4,** 73–131.

CHATTAWAY, M. M. (1933). Ray development in the Sterculiaceae. *Forestry* **7,** 93–108.

— (1938). The wood anatomy of the family Sterculiaceae. *Phil. Trans. B.* **228,** 313–65.

CHEADLE, V. I. and ESAU, K. (1964). Secondary phloem of *Liriodendron tulipifera. Univ. Calif. Publ. Bot.* **36,** 143–252.

CHRYSLER, M. A. (1913). The origin of the erect cells in the phloem of the Abietineae. *Bot. Gaz.* **56,** 36–50.

CUMBIE, B. G. (1963). The vascular cambium and xylem development in *Hibiscus lasiocarpus. Am. J. Botany* **50,** 944–51.

— (1967). Developmental changes in the vascular cambium of *Leitneria floridana. Am. J. Botany* **54,** 414–24.

— (1969). Developmental changes in the vascular cambium of *Polygonum lapathifolium. Am. J. Botany* **56,** 139–46.

DODD, J. D. (1948). On the shapes of cells in the cambial zone of *Pinus sylvestris* L. *Am. J. Botany* **35,** 666–82.

ESAU, K. (1938). Ontogeny and structure of the phloem of tobacco. *Hilgardia* **11,** 343–424.

EVERT, R. F. (1959). Ray origin in *Pyrus communis* L. *Cong. Int. Bot. 9th* **2**, 108.

— (1961). Some aspects of cambial development in *Pyrus communis*. *Am. J. Botany* **48**, 479–88.

KLINKEN, J. (1914). Ueber das gleitende Wachstum der Initialen im Kambium der Koniferen und den Markstrahlenverlauf in ihrer sekundären Rinde. *Bibl. Bot.* **19**, 1–37.

KRIBS, D. A. (1935). Salient lines of structural specialization in the wood rays of dicotyledons. *Botan. Gaz.* **96**, 547–57.

SCHMIDT, E. (1889). Ein Beitrag zur Kenntnis der secondären Markstrahlen. *Ber. dt. bot. Ges.* **7**, 143–51.

SRIVASTAVA, L. M. (1963a). Secondary phloem in the Pinaceae. *Univ. Calif. Publ. Bot.* **36**, 1–142.

— (1963b). Cambium and vascular derivatives of *Ginkgo biloba*. *J. Arnold Arbor.* **44**, 165–92.

THOMSON, R. B. (1914). On the comparative anatomy and affinities of the Araucarineae. *Phil. Trans. B.* **204**, 1–50.

WHALLEY, B. E. (1950). Increase in girth of the cambium in *Thuja occidentalis* L. *Can. J. Res. C.* **28**, 331–40.

4
Variations in the size of fusiform cambial initials

Most of our present information on changes occurring in the vascular cambium of plants has been obtained from investigations made on the secondary xylem and secondary phloem, and not from direct examinations of the cambium itself. Since the fusiform initials leave a record of their activities in both the secondary xylem and secondary phloem, either tissue can be used to study the sequence of changes in cell size and arrangement in the meristem. For a time there was a tendency to use the phloem derivatives for studying changes that had occurred in the cambium, the most notable piece of research being that by Klinken (1914), who, using the phloem derivatives of *Taxus baccata*, first developed the technique of examining serial tangential sections for studying cell changes in the cambium. One reason for using the phloem in preference to the xylem is that the phloem elements change less in size during maturation, and hence the tangential sections were thought to give a more accurate picture of the probable lengths of the cambial initials at the time these cells were formed. However, Esau and Cheadle (1955) have shown that divisions, varying in plane from vertical to almost transverse, are common within the phloem derivatives of the vascular cambium. The derivatives of a division that is nearly transverse will obviously be shorter than the fusiform initial from which they are indirectly derived. Since some of these derivatives will differentiate into sieve elements which do not elongate further before maturation, the length of the sieve elements may have no relation to the length of the fusiform initials. Secondly, as other recent investigators have pointed out, if the sequence of divisions and size changes in the cambium is to be followed over any prolonged period of time the secondary xylem must be used, since distortion and abscission affect the phloem. Furthermore, since divisions are more frequent to the xylem side of the cambium, this tissue gives a more complete picture of events. Bannan (1950) points out that some anticlinal divisions in the cambium, especially those followed by the loss of certain cells, are not recorded in the phloem.

Data on changes in the cambium obtained from investigations on the xylem and phloem and not on the cambium itself, however, do not give us a complete or entirely accurate picture of the activities of this meristem. Although a record of the principal size changes and divisions in the cambium is preserved in its derivative tissues, data obtained from derivative elements of necessity are erroneous to the extent of the elongation of the matured elements. Regrettably few results have been published on the relation in size between the cambial cells and their derived elements. Bailey (1920), from a comparative study of the size variations of cambial initials and their derivatives, concluded that in *Ginkgo* and the Coniferae the length of the tracheids of the last-formed xylem closely resembled that of the fusiform initials from which they were derived but were slightly longer. He concluded that the elongation occurring during differentiation was of the order of 5–10%. In dicotyledons he found that the initials were much shorter than the fibres but were approximately the same length as the vessel members. Chattaway (1936) found that the fibres in dicotyledonous woods were 1·1–9·5 times longer than the cambial initials, though elongation to several times the original length occurs only when the initials are very short.

The assumption that the patterns of variation in length of xylem elements reflect the principal size changes in the cambial initials can hold therefore for only two cell types. Since conifer tracheids elongate relatively little during differentiation, variations in their length may be taken as an approximate indication of the size fluctuations in the cambium. Dicotyledonous vessel members also change very little in length during differentiation. On the other hand, variations in the length of dicotyledon fibres are due in part to intrusive growth during differentiation. Studies in plants with storeyed cambia, where the fusiform initials remain approximately the same length throughout the growing season, have shown that fibres elongate by varying amounts, depending on their position within the growth ring. Factors such as temperature influencing cell-wall plasticity may produce changes in the degree of fibre elongation. Hence the patterns of size changes in wood fibres cannot be regarded as indicating the size changes that have occurred in the cambium. The method of measuring the vessel elements is important if these cells are to be used as a measure of the length of the fusiform initials at the time they were cut off from the cambium. Chalk and Chattaway (1934) concluded that the 'total member length' as measured from tip to tip of the cell corresponded more closely in length with the fusiform initial than did 'mean body length' as measured from the centre of the pores and ignoring any tail or extension. Philipson and Butterfield (unpublished data) found in the wood of *Hoheria angustifolia* Raoul, a plant with a storeyed cambium, that the mean body length of the vessel members of the last-formed xylem corresponded more closely with the mean storey height of the fusiform initials than did the total vessel-member length. Minor fluctuations in the total vessel-member length were found to exist within the growth ring, although there are no comparable changes in the size of the fusiform initials in a storeyed cambium. These fluctuations are due to the broad vessel elements of the early wood having almost transverse end walls compared with the more oblique end walls of the narrow elements formed in the late wood.

Size variations within the one tree

The general pattern of variation in tracheid size within the secondary xylem of conifers was established by Sanio (1872) for *Pinus sylvestris*, and has since been observed in many other softwoods and hardwoods. The literature dealing with the variation in size of the xylem elements has been reviewed by Spurr and Hyvarinen (1954) and Dinwoodie (1961). Two major patterns of variation occur which may be thought of as: (1) horizontal, or occurring across the growth rings from the primary xylem outwards, and (2) vertical, or the pattern occurring up the tree within any given growth ring. These two aspects of variation in initial length will be considered separately along with the other factors that superimpose minor fluctuations on the basic patterns.

Changes in the fusiform cambial cell length at any one level within the tree

From his observations on *Pinus sylvestris*, Sanio concluded that tracheid length at any one level in the tree increased outwards from the pith through a number of annual rings until a definite size was reached, which then remained constant through the following rings (Fig. 4.1). While there has since been

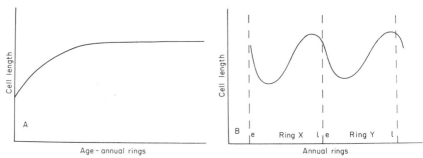

Fig. 4.1 Graph A illustrates the generalized pattern of increase in the average length of derivatives in passing from the innermost to the outermost secondary xylem of the stem. Graph B illustrates this variation within individual growth rings, (e) early wood and (l) late wood of two adjacent growth rings.

general agreement (Lee and Smith, 1916; Anderson, 1951; Hejnowicz and Hejnowicz, 1956) that there is an initial increase in tracheid length over the first growth rings until a maximum is reached, investigations beyond this point have been rather contradictory. Some investigators have recorded results similar to those of Sanio, while others record a further slight increase or even a decrease in mean cell length. Most investigators have observed minor fluctuations in the mean cell length in the outer growth rings, but these are mainly due to secondary influences which will be discussed later.

Once a group of cambial initials has been laid down at the end of primary growth it will move horizontally outwards as the stem grows radially. Radial growth is accompanied by circumferential expansion of the cambium which involves an increase in the number of fusiform cells in the cambium (Chapter 2). This occurs in the non-storeyed cambium by means of a pseudotransverse wall being laid down usually near the centre of the dividing cell. The two daughter initials so formed then elongate by the intrusive growth of their cell tips. The resultant size of the fusiform initials in the cambium is governed by changes in the frequency of pseudotransverse division and the rate and amount of elongation of new initials.

(a) *Frequency of pseudotransverse division.* The length of the cambial fusiform initials is related in part to the frequency of the anticlinal divisions in the cambium owing to the semi-transverse nature of these divisions. A high frequency of pseudotransverse division naturally tends to depress cell length, whereas a low frequency is usually associated with a high mean cell length.

Although high frequencies of pseudotransverse division in the cambium tend to depress cell length, the resultant reduction is not as great as might be expected, since high rates of anticlinal division are usually accompanied by the loss of initials and more rapid elongation of the surviving daughter initials.

(b) *Elongation of new cambial initials.* The rate and amount of elongation of new fusiform cambial initials is naturally an important factor governing the size of the cells at any particular time. For example, the rate of fusiform cell elongation following division varies widely according to the time of year. Bannan (1951b) found that in both *Chamaecyparis* and *Thuja* the rate of elongation of sister fusiform initials was slow during the first quarter of the year's growth, tending to increase during the second and third quarters of the year and reaching a maximum during the last quarter. Evert (1961), however, found in *Pyrus communis* that most elongation occurred before the cambium had resumed activity each year. Local factors often operate to produce irregularities in the rate of cell elongation. For example, after the loss of an initial a neighbouring initial may elongate rapidly into its place. Since this loss of initials is highest towards the end of the year's growth, it no doubt plays an important part in the increase in cell elongation during this period. This results in a high average cambial cell length towards the end of the growth ring. There is considerable variation in the rate and amount of cell elongation from cell to cell in the cambium. One fusiform initial may elongate a great deal and another only slightly. The presence of obstructing rays may reduce the possible elongation of new initials.

(c) *Preferential loss of new fusiform initials.* A third factor that plays a part in the maintenance of a high mean cell length is the preferential loss of short initials after pseudotransverse division. Loss of initials occurs in various ways in all parts of the tree (Bannan, 1951a). The pitch and length of the anticlinal wall at the time of pseudotransverse division fluctuate widely. This can result in two long daughter initials lying almost side by side where the anticlinal wall is very oblique, or in the case where the division is almost transverse the new cells are comparatively short. Often an uneven division results in the loss of the shorter daughter initial. Progressive shortening occurs, resulting in a contraction in size of the radial file of derivatives produced by a disappearing initial. After the pseudotransverse division the initial usually undergoes sporadic periclinal divisions, progressively shortening, and then ceases division altogether. Initials which are lost from the cambium lose their capacity for periclinal division and pass off into maturation, becoming malformed xylem or phloem elements or undergoing reduction to become ray initials. The continuation of the longest cells and the elimination of the shortest is a factor in the maintenance of an efficient cell length (Bannan, 1957b).

The interplay of these factors, (*i*) the frequency of the pseudotransverse divisions, (*ii*) the amount and rate of elongation of the new cells, and (*iii*) the preferential loss of the shorter daughter initials, influences the mean cell size at any given distance from the centre of the stem. Over the first few centimetres of radial growth the perimeter of the stem is increasing very rapidly, and a much higher rate of anticlinal division is required to provide new initials for the cambium than is required later in radial growth. Consider a hypothetical case where the cambium is thought of as a cylinder of ever-increasing diameter. Since the circumference of this cambium is always proportional to its radius ($C = 2\pi r$), for a 1 cm increment of radial growth occurring from 1 to 2 cm radius, the cylinder will have doubled its circumference, i.e. during this 1 cm of radial growth every fusiform initial of the cambium will have on the average divided anticlinally at least once. This is assuming that the daughter initials elongate to the length of the parent initial and retain the same tangential dimension. However, farther out, at, say, 20 cm from the stem centre, for an equal increment of radial growth, the circumference will increase by only one twentieth, i.e. during this 1 cm of radial growth from 20 to 21 cm radius only one cell in every twenty will need to divide anticlinally. The 'required rate' of anticlinal division per centimetre of radial growth which is necessary to maintain the cambium is illustrated graphically in Fig. 4.2, curve B.

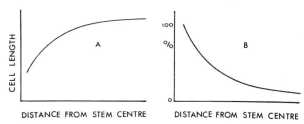

Fig. 4.2 Curve A illustrates the generalized pattern of increase in cell length with radial distance from the stem centre common to most plants with non-storeyed cambia. Curve B illustrates the decrease in the percentage of the cells of the cambial cylinder which must divide anticlinally for every increment of radial growth.

It is evident, therefore, that a lower frequency of anticlinal division in the cells of the cambial cylinder is required when it has attained a greater diameter than during its early growth. In fact, over the outer portion of radial stem growth the rate of anticlinal division is higher than is required to supply the needs of the expanding cambium, and a high loss of initials occurs. Over the first few centimetres distance from the stem centre, although the rate of anticlinal division is relatively high (Bannan, 1950) and the loss of initials somewhat lower than farther out, the resultant effective supply of new fusiform initials does not alone supply the needs of the rapidly expanding perimeter of the cambium. During this period the 'deficiency' of initials is compensated for by an increase in the dimensions of the daughter initials beyond those of their parents. While some increase in the tangential dimension of initials occurs, this is of less importance than their elongation well beyond the length of the former initials. That is to say, a steady increase in cell length occurs outwards during the early growth of a stem. This increase in length makes an important contribution to the circumferential increase of the cambium (Bailey, 1923), because the number of cells at any level becomes increased by intrusion of adjacent cells. These considerations explain the nature of the curve shown in Fig. 4.2A, which is typical of most trees with a non-storeyed cambium.

In the above hypothetical case the rate at which new daughter initials reach or exceed the length of their former parent cell was not considered. The very high rate at which the daughter cells attain their full size during the first few centimetres of radial growth further adds to the maintenance of the cambium during this period of rapid expansion of the stem. It must be emphasized,

however, that all these processes are complementary, and one is not necessarily the cause of the other. In plants with non-storeyed cambia, where no
elongation of new initials occurs after pseudotransverse division, there is
necessarily a decrease in the mean cell length outwards from the centre of the
stem. An interesting example where this type of behaviour occurs is found in
Hibiscus lasiocarpus. In this species Cumbie (1963) found that the anticlinal
divisions occurred in many planes, varying from pseudotransverse to truly
radial longitudinal. This partly compensates for the lack of elongation in new
cells. In the storeyed cambium, where all the anticlinal divisions are in the
radial longitudinal plane, the percentage rate of anticlinal divisions per centimetre of radial growth is sufficient to meet the requirements of the expanding
cambium, except possibly in the first few millimetres of radial growth. In this
region, adjacent to the primary xylem, many woods that otherwise possess a
typically storeyed nature in the outer regions of the stem show a somewhat
non-storeyed appearance.

Secondary influences

While the decreasing rate of necessary anticlinal division in the cambium is
the primary cause of the increase in mean fusiform initial length for a distance
outwards from the centre of the stem, secondary influences may impart
fluctuations to the basic pattern or changes in the maximum cell length
attained or the rate at which it is attained. The most important of these factors
is the actual rate of radial growth with time as expressed by growth-ring
width. There are numerous reports on the relation of tracheid length to ring
width (see Dinwoodie, 1963). Although some authors have claimed that wide
rings induce the formation of long cells, the converse is generally the case.
Bisset, Dadswell, and Wardrop (1951) showed that, in general, wide rings
correspond with short tracheid length in *Pinus radiata*. Other authors have
been unable to find any relationship.

   The problem of the relationship between ring width and cambial behaviour
with reference to cell length was explored by Bannan (1959) by examining
white cedar stems of the following types: (1) stems of similar size but with
rings of diverse widths; (2) stems showing a transition in their peripheral
growth from wide rings to narrow rings; (3) stems exhibiting an alteration
from narrow to wide rings, in their final growth; and (4) stems with rings of
varying widths in different sectors. In the first type much variation occurred
from tree to tree, with no consistent relationship between the frequency of
pseudotransverse division in the cambium and ring width. However, mean
cell length at pseudotransverse division showed an increase with decrease in
ring width. In the second type the transition was usually accompanied by a
slight increase in cell length and the third by a small decrease in cell length.
All comparisons revealed a definite tendency for greater cell length to be

associated with decreased ring width. The lack of a relationship between the rate of multiplicative division and ring width, however, led Bannan to conclude that cell size is due not so much to the frequency of pseudotransverse division as to inherent determiners, the latter being influenced to some extent by growth rates.

Further information on the relation of cell length to growth-ring width has been revealed by Bannan in a more recent series of papers. It was shown for *Thuja* (Bannan, 1960) that, among trees of similar diameter, the maximum cell lengths occurred in growth rings less than 1 mm wide, while trees with wider growth rings tended to have shorter cells. However, the rate of pseudotransverse division was uniform in both fast-grown trees (i.e. those with wider growth rings) and slow-grown trees (i.e. those with narrower growth rings). In *Pinus* (Bannan, 1962) maximum cell lengths were also associated with growth rings less than 1 mm wide, but the frequency of pseudotransverse division increased sharply as ring width decreased below 1·3 mm. Results similar to those for *Pinus* were found in *Picea* (Bannan, 1963a) and *Cupressus* (Bannan, 1963b). This increase in the rate of pseudotransverse divisions tends to decrease the cambial cell length in trees with very narrow growth rings. Hence, although there is, in general, an inverse relationship between ring width and cambial cell length, the sharp increase in the rate of pseudotransverse division in very narrow rings can counteract this effect by tending to depress cell length. Although this inverse relationship holds, Bannan (1965) found that the yearly amount of cambial cell elongation dropped as ring width decreased, but not in proportion to the decline in the ring width. Thus, the cumulative elongation through a lineal series of cells per centimetre of xylem increment increases as ring width decreases, resulting in the amount of elongation occurring during the production of several narrow rings greatly exceeding that through a single ring having the same width. The higher rate of pseudotransverse division towards the end of the growth ring naturally lowers the mean length of the cambial cells at this stage. Bannan (1954) reported that the pseudotransverse divisions sometimes occur earlier in the elongation cycle in more vigorously growing trees than in slower-growing ones of similar diameter. This may cause a slight decrease in cell length in rapidly growing trees.

Chalk and Ortiz (1961) have suggested that the decrease in mean cell length in wide rings in *Pinus radiata* is due to an increase in the number of pseudotransverse divisions occurring in the xylem mother cells. Bannan (1957a), however, has shown that extra-initial pseudotransverse divisions are very few and normally constitute only about 2% of this type of division, though in a very fast grown specimen of *Cedrus deodora* (Roxb.) Lovd. a figure of 20% was recorded. Chalk and Ortiz believe that this increase in the number of pseudotransverse divisions in the xylem mother cells in the early wood is a major factor in lowering the length of the derivatives in this part of the growth ring.

Although the basic pattern for increase in cell length outwards from the centre of the stem, at any one level, has been established by numerous workers for equally numerous trees, remarkably few investigations have been made of the relative effects of age and distance from the pith on tracheid length. Anderson (1951) in a study of a number of conifers, found that cell length at a given distance from the pith was the same throughout the trunk, irrespective of the number of rings involved. Hejnowicz and Hejnowicz (1958), from measurements made on a 50-yr-old tree of *Populus tremula* L., obtained results similar to those reported by Anderson for conifers; the length of elements at a given distance from the pith was the same throughout the trunk irrespective of the number of rings involved, except in this instance for the lower part of the trunk up to 2 m above ground. At the same time they concluded that the distance from the pith was a comparative measure of the number of cambial generations involved in wood formation, i.e. a measure of the relative age of the cambium. This meant that elements situated at the same distance from the pith at various levels were formed by the cambium at the same relative age. Thus the length of the vessel elements, and by inference the fusiform initial length, is more closely related to the relative age of the cambium than to the absolute age of the cambium as measured in terms of growth rings.

It would appear that in the early part of radial growth, where the ratio of anticlinal to periclinal divisions in the cambium is very high, the actual radial distance from the stem centre is of prime importance in governing the balance between the various factors influencing cell size in the cambium. In the region more distant from the stem centre, however, where the frequency of anticlinal divisions has diminished and loss of initials is high, secondary factors, such as growth-ring width, may influence the size of the initials. This would account for the different patterns obtained by various workers for cell length in the later part of the curve, i.e. in the wood farther removed from the stem centre. For example, a decrease in cell length in this later region, as recorded by many workers, could be due to the cumulative influence of several wider growth rings during this period of growth. Conversely, a number of narrow rings could cause a slight increase in the mean cell length. The actual radial distance from the centre of the stem at which this transition in influence occurs will naturally vary from one species to another, its position being influenced by the actual rates of anticlinal division and the percentage loss of initials, these factors being inherent in a particular species. It is a logical conclusion, therefore, that the diameter of the cambium at its initiation will influence the pattern of variation in cell length outwards from the centre of the stem. In plants where the cambium develops outside a large core of primary tissue the characteristic increase in cell length may be partially or totally absent. In such cases it must be assumed that the diameter at its initiation is such that the cambium is able to divide anticlinally at a rate sufficient to provide an adequate supply of new cells. The new cells meet the needs of the rapidly expanding cambial cylinder, and the increase in length of cells is suppressed.

Variations in length of fusiform initials within the growth ring

There are no published results of direct cambial measurements taken at intervals during the growing season, so most of our knowledge on within-ring variation comes from the many studies on the size of the cambial derivatives. Early wood elements are shorter than those produced later in the same annual ring (Fig. 4.1B). This refers to fibres, tracheids (Lee and Smith, 1916; Kribs, 1928; Bisset, Dadswell, and Amos, 1950; Bosshard, 1951; Dinwoodie, 1963; and others), and vessel members in non-storeyed woods (Chalk and Chattaway, 1935; Bisset and Dadswell, 1950; Bosshard, 1951).

Anticlinal divisions in the fusiform initials occur predominantly at the end of the growing season (Bannan, 1950) and are partly responsible for the cyclic changes in the lengths of their derivatives. Elongation of daughter fusiform initials is also uneven, elongation reaching a maximum in the last quarter of the growing season. Part of this variation within the ring, however, is due to the elongation of the derivative cells after they have been cut off from the cambium. This is particularly so in the case of wood fibres and fibre tracheids. Some light has been thrown on the degree and timing of this intrusive growth of derivatives by studies on plants with storeyed cambia (Chapter 5). In these plants anticlinal divisions in the cambium are all in the radial longitudinal plane, and no increase in length of the cambial cells occurs outwards from the pith. Within the growth ring, fibre length rises to a maximum in the middle of the ring and then drops abruptly to the ring boundary (Chalk, Marstrand, and Walsh, 1955). Vessel element and parenchyma strand lengths remain constant throughout the growth ring.

Changes in the fusiform cambial initial length in relation to height in the tree

There is a gradient in cell length along the axis of the tree. When the pattern of cell length is followed upwards in the same growth ring the average length increases from the base upwards, reaching a maximum and then decreasing towards the top (Sanio, 1872; Pritchard and Bailey, 1916; Bailey and Shepard, 1915; Lee and Smith, 1916; Chalk, 1930; Bethel, 1941; Bisset and Dadswell, 1949), the longest cells occurring at a higher level in successive annual rings (Bailey and Shepard, 1915; Bethel, 1941).

With increasing height up the stem, within any one growth ring, the cell length increases from the base to about one-third of the stem height and then decreases. Thus the point of maximum length is located at progressively higher levels as successive annual rings are formed (Fig. 4.3A). Chalk (1930) pointed out that the different patterns of variation in cell size within the tree could be different aspects of the same phenomenon. If this is so, the upward gradient is a result of the fact that the radial distance of the cambium from the centre of the stem decreases with the tapering of the tree. While this is in part true, its influence is most pronounced towards the very top of the trunk, where the stem diameter is relatively small. The changing radial distance of the cambium from the stem centre cannot explain the increase in cell length over the first third of the total stem height. Hejnowicz and Hejnowicz (1958) explained the pattern in this region of the stem as influenced by its proximity to the root.

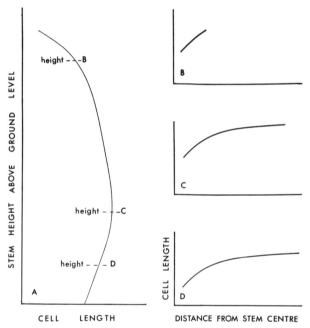

Fig. 4.3 Graph A illustrates the typical pattern of change in cell length with increasing height up the stem. Graphs B, C, and D show the increase in cell length with radial distance from the stem centre at the corresponding heights B, C, and D indicated on graph A.

Another approach to the problem of variation in cell length with tree height comes from our knowledge that trees pass through a grand period of growth before entering old age. It would appear certain that the rate of height growth at a particular time has an influence on the size of the cambial initials originating during that period. Dinwoodie (1963) found evidence that rings laid down by the cambium in short internodes possessed shorter tracheids than did rings laid down in long internodes. Where very little stem elongation has occurred, the cambial initials laid down by the apical meristem are apparently shorter than those occurring as a result of normal growth. This relation between amount of height growth and initial length has its greatest effect in the first growth rings. Mell (1910), Lee and Smith (1916), and others also noted that faster-growing trees possess longer tracheids, but did not correlate this with internode length. Trees with this high initial cambial cell length will usually maintain the relatively high length throughout subsequent growth rings (Bisset, Dadswell, and Wardrop, 1951). Many trees pass through a grand period of growth when the actual rates of height and diameter growth are at a maximum. It would be expected therefore that initials laid down during this period might be appreciably longer than those laid down at periods of slower growth. The general pattern of increase then decrease in cell length when followed in any one growth ring up the tree could thus be partly explained in terms of height growth (Fig. 4.3B, C, D). Towards the top, however, when the stem diameter becomes relatively small, the decrease in cell size is due to the 'core' effect, i.e. the increase in cell length outwards from the stem centre has not reached its maximum in this region (Fig. 4.3B).

Secondary influences

Just as secondary influences impart fluctuations to the basic pattern of cell size outwards from the stem centre at any one level, so, too, do secondary influences cause fluctuations to the normal pattern of cell size with tree height. The most prominent of these is the rate of height growth. If, for any environmental or other reason, the rate of height growth is suppressed in any one year the mean cell size of the new cambial initials laid down during this period will be shorter. Even after radial growth of several years these initials will be shorter than those in neighbouring internodes laid down during more favourable conditions of height growth.

Other causes of variation in cell size

Kaeiser (1964) has shown in Eastern cottonwood that the fusiform cambial initials show an increase in length on both the upper and lower sides of a leaning trunk. Generally, however, compression wood is characterized by shorter tracheids than the corresponding normal wood (Dadswell and Wardrop, 1949; Bisset and Dadswell, 1950). Some workers have shown that tension-wood fibres are longer than those in normal wood. It is not clear if these size changes in fact occur in the cambium, or are the result of reduced or increased amounts of intrusive elongation of the derivatives.

The pattern of size changes within the internode has not as yet been adequately investigated. Bailey and Tupper (1918) noted that tracheids tended to be shorter at nodes, the junction of stems and roots or branches, and other regions where growth adjustments occur. Initials also tend to be shorter at the sites of injuries.

Another pattern of variation which departs from the general pattern for trees has been recorded by Rumball (1963). This is found in trees which pass through a juvenile stage marked by a striking change in habit to that of the adult tree. The rate of increase in cell length outwards along any one radius was found to be very rapid during the juvenile phase, but when the tree entered the adult phase the rate of increase declined after a sharp break in the graph representing cell length against distance from the pith. This break in the curve corresponds with the time of the habit change from a divaricate shrub to a normal tree.

Size variation between different trees

Differences in cell length also occur between specimens of the same species with different provenances. Echols (1958) found a significant relationship between tracheid length and latitude in *Pinus sylvestris* when plants from different latitudes were grown together. Cell length showed a decrease with increasing latitude. A similar relationship was observed by Dinwoodie (1963) for Sitka spruce. However, the variation in cell length between provenances is small (5–30%) compared with the differences in cell length which occur within any one tree (e.g. a 200–400% increase in cell length outwards from the pith).

Harlow (1927), working with *Thuja occidentalis*, concluded that the differences in site were negligible in view of the fluctuations among trees growing on the same site. Bannan (1959) examined nearly 100 trees from diverse habitats and corroborated Harlow's findings. Site would seem important only in as much as it affects growth generally. Bannan (1963b) studied the interaction of environmental and inherent factors on cambial cell size in *Cupressus*. In general, he found the species from arid regions had smaller cells than those from the damper regions. The effects of temperature and day length on tracheid size have been studied by other workers, but since no measurements have ever been made directly on the cambial initials, it is not clear how much these factors influence the size of the initials or how much they alter the rate and amount of elongation of the derivatives by changing the cell-wall plasticity or otherwise affecting differentiation.

REFERENCES

ANDERSON, E. A. (1951). Tracheid length variation in conifers as related to distance from the pith. *Forestry* **49**, 38–42.

BAILEY, I. W. (1920). The cambium and its derivative tissues. II. Size variations in cambial initials. *Am. J. Botany* **7**, 355–67.

— (1923). The cambium and its derivative tissues. IV. The increase in girth of the cambium. *Am. J. Botany* **10**, 499–509.

— and SHEPARD, H. B. (1915). Sanio's laws for the variation in size of conifer tracheids. *Botan. Gaz.* **60**, 66–71.

— and TUPPER, W. W. (1918). Size variation in tracheary cells. I. A comparison between the secondary xylems of vascular cryptogams, gymnosperms and angiosperms. *Proc. Am. Acad. Arts Sci.* **54**, 149–204.

BANNAN, M. W. (1950). The frequency of anticlinal divisions in fusiform cells of *Chamaecyparis*. *Am. J. Botany* **37**, 511–19.

— (1951a). The reduction of fusiform cambial cells in *Chamaecyparis* and *Thuja*. *Can. J. Botany* **29**, 57–67.

— (1951b). The annual cycle of size changes in the fusiform cambial initials of *Chamaecyparis* and *Thuja*. *Can. J. Botany* **29**, 421–37.

— (1954). Ring width, tracheid size and ray volume in stem wood of *Thuja occidentalis* L. *Can. J. Botany* **32**, 466–79.

— (1957a). The relative frequency of the different types of anticlinal division in conifer cambium. *Can. J. Botany* **35**, 425–34.

— (1957b). The structure and growth of the cambium. *TAPPI* **40**, 220–5.

— (1959). Some factors influencing cell size in conifer cambium. *Proc. 9th Int. Bot. Congress*, 1704–9.

— (1960). Cambial behavior with reference to cell length and ring width in *Thuja*. *Can. J. Botany* **38**, 177–83.

— (1962). Cambial behavior with reference to cell length and ring width in *Pinus strobus* L. *Can. J. Botany* **40**, 1057–62.

— (1963a). Cambial behavior with reference to cell length and ring width in *Picea*. *Can. J. Botany* **41**, 811–22.

— (1963b). Tracheid size and anticlinal division in the cambium of *Cupressus*. *Can. J. Botany* **41**, 1187–97.

— (1965). The rate of elongation of fusiform initials in the cambium of the Pinaceae. *Can. J. Botany* **43**, 429–35.

BETHEL, J. S. (1941). The effect of position within the bole upon fibre length of loblolly pine (*Pinus taeda*). *J. Forestry* **39**, 30–3.

BISSET, I. J. W. and DADSWELL, H. E. (1949). The variation of fibre length within one tree of *Eucalyptus regnans*. *Australian For.* **13**, 86–96.

— and — (1950). Variation in cell length within the growth ring of certain angiosperms and gymnosperms. *Australian For.* **14**, 17–29.

—, — and AMOS, G. L. (1950). Changes in fibre length within one growth ring of certain angiosperms. *Nature, Lond.* **165**, 348–9.

—, — and WARDROP, A. B. (1951). Factors influencing tracheid length in conifer stems. *Australian For.* **15**, 17–30.

BOSSHARD, H. H. (1951). Variabilitat der Elemente der Eschenholzes in Funktion von der Kambiumtatigkeit. *Schweiz. Z. Forestw.* **102**, 648–65.

CHALK, L. (1930). Tracheid length with special reference to Sitka spruce (*Picea sitchensis* Carr.). *Forestry* **4**, 7–14.

— and CHATTAWAY, M. M. (1934). Measuring the length of vessel members. *Trop. Woods* **40**, 19–26.

— and — (1935). Factors affecting dimensional variations of vessel members. *Trop. Woods* **41**, 17–37.

— and ORTIZ, M. (1961). Variation in tracheid length within the ring of *Pinus radiata* D. Don. *Forestry* **34**, 119–24.

—, MARSTRAND, E. B. and WALSH, J. P. (1955). Fibre length in storeyed hardwoods. *Acta Bot. Neerl.* **4**, 339–47.

CHATTAWAY, M. M. (1936). The relation between fibre and cambial initial length in dicotyledonous woods. *Trop. Woods* **46**, 16–20.

CUMBIE, B. G. (1963). Vascular cambium and xylem development in *Hibiscus lasiocarpus. Am. J. Botany* **50**, 944–51.

DADSWELL, H. E. and WARDROP, A. B. (1949). What is reaction wood? *Australian For.* **13**, 22–33.

DINWOODIE, J. M. (1961). Tracheid and fibre length in timber. A review of literature. *Forestry* **34**, 125–44.

— (1963). Variation in tracheid length in *Picea sitchensis* Carr. *Forest Products Special Report 16 D.S.I.R.* (H.M.S.O. London).

ECHOLS, R. M. (1958). Variation in tracheid length and wood density in geographic races of Scotch pine. *Yale Univ. Sch. For. Bull.* **64**.

ESAU, K. and CHEADLE, V. I. (1955). Significance of cell divisions in differentiating secondary phloem. *Acta Bot. Neerl.* **4**, 348–57.

EVERT, R. F. (1961). Some aspects of cambial development in *Pyrus communis. Am. J. Botany* **48**, 479–88.

HARLOW, W. H. (1927). The effect of site on the structure and growth of white cedar (*Thuja occidentalis* L.). *Ecology* **8**, 450–70.

HEJNOWICZ, A. and HEJNOWICZ, Z. (1956). Badania anatomiczna nad drewnem topoli. *Arboretum Kornickie* **2**, 195–218. (In Polish with English summary.)

— and — (1958). Variation of length of vessel members and fibres in the trunk of *Populus tremula* L. *Acta Soc. Bot. Pol.* **27**, 131–59.

KAEISER, M. (1964). Vascular cambium initials in Eastern Cottonwood in relation to mature wood cells derived from them. *Trans. Illinois Acad.* **57**, 182–4.

KLINKEN, J. (1914). Ueber das gleitende Wachstum der Initialen im Kambium der Koniferen und der Markstrahlenverlauf in ihrer sekundaren Rinde. *Bibl. Bot.* **19**, 1–37.

KRIBS, D. A. (1928). Length of tracheids in jack pine in relation to their position in the vertical and horizontal axis of the tree. *Minn. Agr. Exp. Sta. Bull.* **54** (not seen).

LEE, H. N. and SMITH, E. M. (1916). Douglas fir fibres with special reference to length. *For. Quarterly* **14**, 671–95.

MELL, C. D. (1910). Determination of quality of locality by fibre length of wood. *For. Quarterly* **8**, 419–22.

PRITCHARD, R. P. and BAILEY, I. W. (1916). The significance of certain variations in the anatomical structure of wood. *For. Quarterly* **14**, 662–70.

RUMBALL, W. (1963). Wood structure in relation to heteroblastism. *Phytomorphology* **13**, 206–14.

SANIO, K. (1872). Ueber die Grosse der Holzellen bei der gemeinen Kiefer (*Pinus sylvestris*). *Jahrb. wiss Bot.* **8**, 401–20.

SPURR, S. H. and HYVARINEN, M. J. (1954). Wood fibre length as related to position in tree and growth. *Botan. Rev.* **20**, 561–75.

5

The storeyed cambium

When viewed in tangential section the cells of the vascular cambium and its derivatives show one of two basic patterns. In gymnosperms and in many dicotyledonous plants the cambial initials are arranged in a more or less irregular pattern with the ends of adjoining cells overlapping. In some of the more highly specialized dicotyledonous plants, however, the cells appear to be in definite horizontal rows or tiers (Fig. 5.1). The cells in such an arrangement are said to be storeyed or stratified.

The storeyed arrangement of the cambial initials may result in a similar stratified condition being present in the secondary xylem and secondary phloem, though in these derivatives the pattern is frequently obscured by the changes that occur during their respective differentiation. This is particularly evident in the case of the xylem, where intrusive growth of the fibres well beyond the limits of the cambial tier may obscure the formerly storeyed pattern.

The fusiform initials in the non-storeyed cambium are vertically elongated cells with tapering ends which overlap with those of vertically adjacent cells (Fig. 1.2). By comparison, the initials of the storeyed cambium are roughly hexangular in outline, with long parallel sides and abruptly tapering ends. The ends of cells in two adjacent storeys do not overlap to any great extent. The fusiform initials of a storeyed cambium are also considerably shorter than those of a non-storeyed cambium. For example, in *Robinia*, a plant possessing a storeyed cambium, the fusiform initials are approximately 175–200 μm long, as compared to those in a conifer, which may have its non-storeyed fusiform initials up to 4,000 μm in length. In an extreme case, such as *Sequoia*, the initials have been known to attain a length of 9,000 μm (Bailey, 1923).

The stratified arrangement of the cells in the storeyed cambium is produced as a direct consequence of the nature of the anticlinal division and the absence of intrusive growth of the daughter cells following such a division. Other

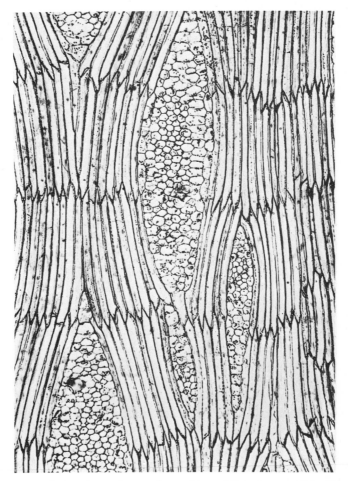

Fig. 5.1 Tangential longitudinal section of the vascular cambium of *Hoheria angustifolia*. The short fusiform initials are in horizontal tiers, with their abruptly tapering tips at approximately the same level. (× 135.)

factors which influence the resulting nature of the cambium include the size of the cambial radius at its initiation and the amount of primary longitudinal growth which continues after radial growth has commenced.

When a stem grows radially the number of cells in the cambium is constantly increasing, thus keeping pace with the increase in the circumference of the stem. In non-storeyed woods this occurs by the pseudotransverse anticlinal division of the fusiform cambial initial followed by the intrusive elongation of the subsequent daughter initials (Chapter 2). Bailey (1923) pointed out, however, that by its very nature, increase in girth of the storeyed cambium cannot involve the elongation of daughter cells to any great extent, or the tierlike nature of the cambium would be lost. In the storeyed cambium the fusiform initials divide anticlinally by the formation of the new cell wall in the radial longitudinal plane (Fig. 2.2). This results in the two daughter initials lying side by side after division, and the absence of any subsequent elongation of these initials permits the development of the characteristic storeyed pattern. As a consequence of this method of division, the number of cells in each storey is continually increasing with increasing radial distance from the stem centre. It is this fundamental difference in the orientation of the new cell wall at anticlinal division, combined with a total absence of intrusive growth of the daughter cells, which separates the storeyed from the non-storeyed cambium.

Compared with the investigations that have been undertaken on non-storeyed cambia, the storeyed cambium has been somewhat neglected. The most comprehensive account published is that of Beijer (1927), who made a meticulous study of cell division in the root cambium of *Aeschynomene elaphroxylon* Taub. (*Herminiera elaphroxylon* Guill. and Perr.). Using the method devised by Klinken (1914), he traced the pattern of cell division through serial tangential sections. One aspect of his investigation involved a study of the readjustments which occur at the cell ends after the anticlinal division. Beijer concluded that the new cell wall may intersect the existing cell wall at any one of a number of positions, but this is always followed by growth adjustments in the form of slight intrusive growth so as to restore the pointed ends to both daughter cells. This limited intrusive growth ensures that the cell walls comply with what he termed 'the law of minimal area'.

The arrangement of the cells at the time of initiation of the cambium in plants later possessing a storeyed cambium has been little studied, possibly due to the difficulty of obtaining good sections of the procambium. Von Hohnel (1884) was of the opinion that the storeyed arrangement of the cells was already present in the procambium and that the development of the storeyed cambium was simply a consequence of this stratification in the procambium. Klinken (1914), however, pointed out that this was not a necessary prerequisite for a storeyed cambium. According to Klinken, the young cambial cylinder expands so rapidly in a tangential direction at the beginning of secondary growth that a distinct storeyed pattern is possible at a very early stage simply by means of the radial longitudinal divisions occurring within the cambial cylinder. Simple calculations on roots of *Aeschynomene* by Beijer (1927) have shown that this is indeed the case. In *Aeschynomene* the cells of the procambium are non-storeyed and the expansion of the initially small cambial cylinder by radial longitudinal divisions results in the development of tiers of cells.

Although it is only in stems of considerable thickness that the tiers of cells in a storeyed cambium assume a marked regularity, the diameter which a stem must attain before the storeyed pattern becomes pronounced is subject to wide variations. Record (1919) reported that, in certain stems of the Leguminosae and Zygophyllaceae, the initials are found to be in regular tiers in what was apparently only the second growth ring and less than $\frac{1}{8}$ in from the pith. In general, however, it is not until stems are considerably thicker that the storeyed pattern is marked.

The apparent variation between species was known to be influenced by growth rates, but it was Beijer (1927) who first realized that the radius of the cambium at the time of its initiation could have an important bearing on the subsequent development of storeys. By contrasting the average number of cells per storey at a given radius in stems and roots where the cambial radius differs at initiation, he concluded that the smaller the radius of the cambium at its initiation, the more cells there would be present in any storey at any given distance from the centre of the stem. Thus, in stems with a large core of pith and primary xylem, the storeyed arrangement of the cells in the cambium will not become apparent until the stem has acquired a considerable thickness, whereas in stems with a small core of primary tissue the tier-like nature of the cambium will become apparent at a much smaller thickness.

Although some morphologists do not recognize the existence of a distinct 'cambium' until primary growth has ended, it is nevertheless true that in some plants radial growth has commenced some time before primary longitudinal extension has ceased. Before a cambium can become storeyed this primary growth must have ended. It is likely that the occurrence of the non-storeyed state near the centre of some stems which later possess a truly storeyed cambium is due, in part, to the occurrence of transverse and pseudotransverse divisions in a meristem that is still being subject to primary extension.

Derivatives of the storeyed cambium remain grouped in horizontal tiers unless elongation of the derivatives modifies this pattern. Usually the storeyed pattern does not remain in all elements. In the extreme case, as in some members of the Zygophyllaceae, the Bignoniaceae and the Leguminosae, all the rays, vessel members, tracheids, fibres, and wood parenchyma strands remain grouped in horizontal tiers. However, this is not the case in most plants with a storeyed cambium. Vessel elements are commonly storeyed and correspond approximately in length to the height of each storey, though occasionally short segments are found that are subdivisions of a storey produced by a transverse division of a mother cell. The parenchyma is also usually storeyed. Cells may be one storey high or form a vertical strand of two, four, or more parenchyma cells within the storey (Fig. 5.2). Record

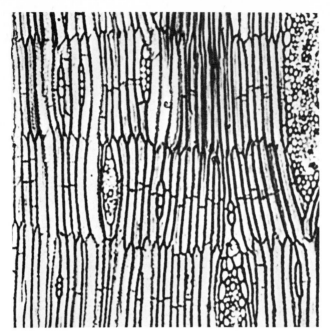

Fig. 5.2 Tangential longitudinal section through the secondary xylem of *Hoheria angustifolia*. The strands of axial parenchyma have either two or four parenchyma cells within the storey, giving the impression of secondary stratifications. (× 150.)

(1919) suggests that it is not uncommon to find metatracheal strands composed of two cells and paratracheal strands of four cells. Where these dividing walls are regularly disposed, they may appear as secondary striations within the storey. Record cited *Bombax, Ceiba,* and *Heliocarpus* as examples having four parenchyma cells per strand, *Charpentiera, Diaphysia,* and *Lonchocarpus* as having two cells per strand, and *Gossypium* and *Pterocymbium* as having two cells per strand in the metatracheal and four cells per strand in the paratracheal parenchyma.

The wood fibres and fibre tracheids more rarely show a storeyed nature, since they seldom remain the same length as the cambial initials from which they are derived. Sometimes the elongated fibres have a widened middle portion corresponding in length to that of the original cambial initial. Where the fibres have undergone no elongation, as in *Triplochiton, Scleroxylon,* and *Aeschynomene,* the cells remain storeyed. In *Aeschynomene* the fibres are similar in size to the parenchyma cells but have thicker walls and more pointed ends (Fig. 5.3).

Rays may be of only one storey in height, e.g. *Daniella thurifera* Bennett, or they may extend through several storeys yet still retain a storeyed pattern.

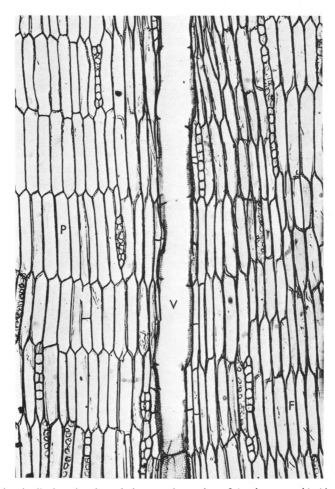

Fig. 5.3 Tangential longitudinal section through the secondary xylem of *Aeschynomene hispida* Willd., a plant where all the derivatives of the storeyed cambium remain in distinct tiers. (P) undivided parenchyma cells which form the bulk of the 'wood', (V) vessel members, and (F) the fibres with thicker walls and slightly more pointed cell tips than the parenchyma cells. (× 100.)

Such an arrangement can give the wood a distinctive pattern of bands or striations when examined macroscopically, termed *ripple marks*. Where the rays are all storeyed they generally occupy the median portion of each tier, the height of which is usually greater than that of the rays. In some storeyed woods the rays may be of two types, the smaller only being storeyed, while the larger are more irregularly dispersed. This is especially common in the Bombacaceae, Malvaceae, and the Sterculiaceae (Record, 1919). There is frequently an aggregation of pits on the fibre end walls in storeyed woods, where the lumina tend to constrict, and this often makes the lines between the storeys more distinct. Solerader (1908) figures a species of *Aeschynomene* with the parenchyma cells pitted prominently at the ends.

Two factors therefore influence the retention of the storeyed pattern in the derivates of a storeyed cambium. These factors are: (*a*) the distribution and size of the ray initials in the cambium, and (*b*) the amount of intrusive growth occurring in the derivatives of fusiform initials. For example, although the fusiform initials might be in clear storeys, the storeyed arrangement is not noticeable because of the presence of many rays of unequal size. Conversely, the storeyed pattern may be visible in the derivatives due to the fact that the rays are nearly all of an even size and lying at the same height, even if the storeyed structure has been lost in the mass of cells due to intrusive growth.

Due to their relative unimportance to the pulp and paper industry, little research has been carried out on the size variations in the cambial derivatives of storeyed hardwoods. Hejnowicz and Hejnowicz (1959) made measurements of fibres and vessel members of *Robinia pseudoacacia* L. They report a distinct change in fibre length within the growth ring, fibre length increasing from the first formed early wood to the late wood, with an abrupt drop in length at the ring boundary. The amplitude of fibre-length variation increased from the pith outwards. They also report that the fibres of the first-formed early wood showed no great tendency to increase from the pith outwards, but the fibres in other parts of the wood showed this tendency as clearly as the fibres in a non-storeyed wood. In contrast to this, Chalk, Marstrand, and Walsh (1955) reported that, in *Pterocarpus angolensis* DC., *Nesogordonia papavifera* Capuron and *Aeschynomene elaphroxylon*, neither the fibres nor the parenchyma strands showed any general tendency to increase in length from the pith outward. Within the growth ring the fibre length rose to a maximum in the middle of the ring and dropped abruptly at the ring boundary, but the length of the parenchyma strands remained constant. Philipson and Butterfield (unpublished data) have shown in *Hoheria angustifolia* that the vessel members within a growth ring display a slight increase in length towards the end of the ring. From associated studies, however, it has been shown that this is due to the angle of inclination of the end wall, the vessel members towards the end of the growth ring being narrower and with more tapering ends compared with the wider transverse-ended vessel members of the early wood. Over seventeen growth rings no increase in the mean length of the vessel members was observed. The possibility that the length of the derivative was dependent upon the plane of the anticlinal division was investigated by Esau and Cheadle (1955), who examined ninety-one species of seventy-one genera representing plants with both storeyed and non-storeyed cambia. They showed in *Asimina triloba* Dunal. (storeyed cambium) that the phloem elements derived from the cambial initials had divided along various planes, resulting in the sieve elements being shorter than the cambial initials from which they were derived. The vessel elements, however, corresponded in length to that of the initials.

Beijer (1927) and others have shown that storeyed structure is related to length of cambial initials, storeyed structure being associated with short initials. It is worth pointing out that some authors have inferred that the change in angle of the anticlinal wall from the pseudotransverse position to the radial longitudinal position has been made possible by the short length of the cambial initials. However, while it is true that storeyed arrangement is associated with short fusiform initials, one is not necessarily the cause of the other. It should be remembered that divisions in the periclinal plane usually occur from tip to tip of the initial irrespective of the length of the cell.

Bailey (1920, 1923) noted from his studies on the length of fusiform initials in various plants that as the initials become shorter, the ends of the more oblique partitions tend to approach the extremities of the cell. Thus certain of the more highly differentiated dicotyledonous plants will tend to show transitional types of meristems with incipient stages of stratifications. Furthermore, he postulated that the variation in size of adjacent fusiform initials in storeyed cambia may be due at least in part to some of the anticlinal divisions being somewhat oblique. The gradation from non-storeyed to storeyed arrangement of cells in some plants has also been noted by Chalk (1937).

The association of storeyed cambia in woods with short vessel elements and other highly specialized features has led to the conclusion that storeyed structure is an advanced character. Storeyed cambia occur in very many systematically widely different dicotyledonous families, perhaps only in some species, or in almost all the members of the family (e.g. the Papilionaceae). This irregular taxonomic distribution of storeyed wood has led to the belief that this character has arisen independently in many lines. In any case, since the development of storeys is dependent on several factors, a simple phylogenetic relationship seems unlikely. Record in a series of papers (1911, 1912, 1919, 1927a, 1927b, 1936, and 1943) has listed families and genera with storeyed wood, giving extensive lists of storey heights and noting which elements are storeyed in a particular wood. Metcalfe and Chalk (1950) also list families where some genera possess storeyed elements. The association of storeyed structure with other advanced features has led taxonomists to use it as a criterion for suggesting that a particular plant is advanced. Metcalfe (1945) concluded that where storeyed structure is confined to certain genera of a family it must be presumed that these genera are more advanced than the remainder.

REFERENCES

BAILEY, I. W. (1920). The cambium and its derivative tissues. III. A reconnaissance of cytological phenomena in the cambium. *Am. J. Botany* **7**, 417–34.

— (1923). The cambium and its derivative tissues. IV. The increase in girth of the cambium. *Am. J. Botany* **10**, 499–509.

BEIJER, J. J. (1927). Die Vermehrung der radialen Reihen im Cambium. *Rec. trav. Bot. Neerl.* **24**, 631–786.

CHALK, L. (1937). The phylogenetic value of certain anatomical features of dicotyledon woods. *Ann. Bot.* **1**, 409–28.

—, MARSTRAND, E. B. and WALSH, J. P. (1955). Fibre length in storeyed hardwoods. *Acta Bot. Neerl.* **4**, 339–47.

ESAU, K. and CHEADLE, V. I. (1955). Significance of cell divisions in differentiating secondary phloem. *Acta Bot. Neerl.* **4**, 348–57.

HEJNOWICZ, A. and HEJNOWICZ, Z. (1959). Variation of vessel members and fibres in the trunk of *Robinia pseudoacacia*. *Proc. 9th Int. Bot. Cong., Abstract Résumes* **2**, 158–9.

HOHNEL, F. VON (1884). Uber stockwerkartig aufgebaute Holzkorper. *Kaiserl. Acad. d. wiss.* **89**, part 1, 30–47.

KLINKEN, J. (1914). Ueber das gleitende Wachstum der Initialen im Kambium der Koniferen und der Markstrahlenverlauf in ihrer sekundaren Rinde. *Bibl. Bot.* **19**, 1–37.

METCALFE, C. R. (1945). Systematic anatomy of the vegetative organs of the angiosperms. *Biol. Rev.* **21**, 159–72.

— and CHALK, L. (1950). *Anatomy of the Dicotyledons*, Oxford University Press, Oxford.

RECORD, S. J. (1911). Tier-like structure of some woods. *Hardwood Rec.* (*Chicago*) **34**, 38–9.

— (1912). Tier-like arrangement of elements in certain woods. *Science, N.Y.* **35**, 75–7.

— (1919). Storied or tier like structure in certain dicotyledonous woods. *Bull. Torrey Bot. Club* **46**, 253–73.

— (1927a). Occurrence of 'ripple marks' in woods. *Trop. Woods* **9**, 13–18.

— (1927b). Note on the ripple marks in the Compositae. *Trop. Woods* **10**, 54.

— (1936). Classification of various anatomical features of dicotyledon woods. *Trop. Woods* **47**, 12–27.

— (1943). Key to the American woods. X. Woods with storied structure. *Trop. Woods* **76**, 32–47.

SOLERADER, H. (1908). *Systematic Anatomy of the Dicotyledons*, English edition, Oxford University Press, Oxford.

6

Modifications to the cambium

In the preceding chapters the normal type of cambium has been considered. A brief account will now be given of modifications of the cambium which occur in different parts of the plant. These regional differentiations are associated with various organs of the plant and give rise to secondary tissues with specialized characteristics. Following this, the next chapter will include a description of plants in which the cambium is found in anomalous positions, or in which it is unusual in some other way. The specialized type of cambium in monocotyledons will be described in Chapter 8.

The normal type so far considered predominates in dicotyledons and gymnosperms. The occurrence of secondary thickening in other plant groups, living and fossil, is reviewed by Barghoorn (1964).

Regional differentiation

Different parts of the cambium within one plant may produce secondary tissues with distinct characteristics. This occurs in several ways.

I. *Distinct fascicular and interfascicular segments*

The cambium may form a complete circle at the time of its first appearance. This will occur if the metaxylem of adjacent primary bundles has united, forming a circle before secondary growth has begun. It is more usual for the primary bundles to remain distinct until secondary thickening begins. In that event the cambium forms first as separate segments within the bundles. These fascicular cambia become linked by interfascicular segments except in stems with very insignificant amounts of secondary thickening.

Whether the cambial ring forms as a whole or is constructed by the linking of separate segments, it is usual for it to produce secondary wood which is

uniform over its whole circumference. That is to say, the tissues derived from
fascicular and interfascicular segments are the same, and this may be so even
when the metaxylem groups are widely separated (Fig. 6.1). As interfascicular
segments of the cambium arise later than the fascicular, and may not produce

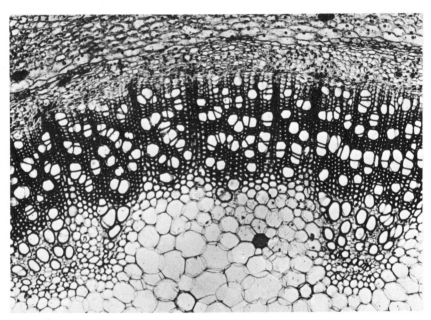

Fig. 6.1 Transverse section of an internode of *Sambucus* at an early stage of secondary
thickening. Large groups of primary xylem occur at each side of the figure and a small group in
the centre. Between these the cambium has laid down secondary xylem similar in all respects to
that opposite the metaxylem. The principal vascular rays abut on the pith. (× 100.)

typical derivatives immediately, there may be some lag in wood production in
these sectors. However, this is normally very transient, so that the tissues
soon become similar all around the axis. Catesson (1964) has described the
establishment of the interfascicular cambium in *Acer pseudoplatanus*, a tree
with narrow medullary rays between the primary bundles.

On the other hand, it is not at all unusual for the production of typical
xylem elements by the interfascicular segments to be delayed, so that sectors
of distinctive secondary tissues are produced. This condition is characteristic
of herbs in which the primary bundles remain separate and in which secondary thickening may be slight. In these stems the derivatives of the interfascicular cambium are typically parenchymatous, though often heavily lignified,
and appear to extend the medullary rays into the secondary tissues. They
differ from vascular rays because they extend over the whole length of the

stem enclosed by a reticulation of the primary vascular system. That is to say they would normally be equal in length to one or more internodes. Another distinctive feature is that they are derivatives of a segment of the cambium and not of a limited area of ray initials.

Such distinctive interfascicular sectors of secondary xylem are not common in woody plants. However, the stems of some species of *Casuarina* illustrate this feature diagrammatically (Fig. 6.2). These sectors are continuous with the pith and correspond to the medullary rays which separate the primary bundles in the young stem. They are continued into the secondary vascular tissue and extend vertically throughout one internode. True vascular

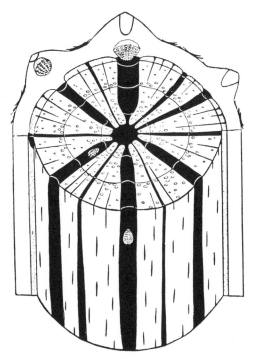

Fig. 6.2 Diagram of the stem of *Casuarina*. The secondary tissues produced by the interfascicular sectors of the cambium are distinctive. (From Jeffrey, 1917.)

rays appear subsequently in the secondary xylem; they do not reach the central pith, nor have they so great a vertical extent. In other species of *Casuarina* the derivatives of the interfascicular cambium eventually come to resemble the remainder of the secondary wood, at least in part, so that these sectors become transformed into normal secondary xylem. These sectors of delayed secondary xylem are sometimes referred to as intermediate bundles.

A similar development of interfascicular sectors extending over a whole

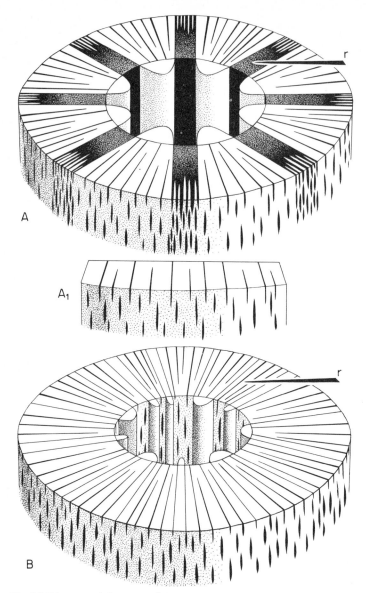

Fig. 6.3 Diagrams of the stems of A. *Fagus* and B. *Sambucus*, much simplified. In *Fagus* the interfascicular derivatives at first are parenchymatous, though they soon become dissected by strands of lignified tissue to form aggregate rays. A₁, the outer surface of the secondary xylem at a later stage, to show loss of the aggregate rays. In *Sambucus* the earliest cambial derivatives are similar over the whole circumference: r; derivatives from cambial sector between primary xylem strands. Compare with Fig. 6.1.

internode is present in *Macropiper* (Balfour, 1958). The so-called aggregate rays of such familiar timbers as the beech, *Fagus sylvatica* L., are derived from similar interfascicular sectors. The primary bundles in the beech are widely separated, and the secondary xylem is also split into segments by wide parenchymatous sectors which correspond to the original medullary rays of the young shoot. These sectors have great vertical extent, and though they soon become divided into segments by local differentiation of fibres, they retain their obvious unity for many years as aggregate rays. Gradually they become subdivided, so that in the timber of a mature tree the vascular rays are not aggregated but are evenly distributed around the circumference.

The distinctive interfascicular sectors just considered must not be confused with a very common situation in which many of the vascular rays, as seen in a transverse section of a young stem, appear to extend from the pith into the secondary xylem (Fig. 6.1). This appearance is due to the vascular rays being formed from the very first derivatives of parts of the interfascicular cambium. If such a stem is viewed in T.L.S. it is evident that the rays are of very limited vertical extent. *Sambucus* has been used to illustrate this very common type of stem in Fig. 6.3, where it is contrasted with the structure of the beech tree. It is recommended that the term primary ray as defined by Frey-Wyssling (1964) be reserved for the type of ray illustrated for *Sambucus*.

Distinctive interfascicular sectors are even more clearly defined in the stems of some lianes (for example, *Clematis* and *Aristolochia*), but it may be more correct to regard these as specialized anomalies, and they will be considered in the next chapter.

II. *Features associated with leaf traces*

The phloem and xylem of the primary bundles become separated by secondary tissues formed by the cambium. Throughout the length of the internode these secondary tissues normally consist of secondary phloem and secondary xylem. In the neighbourhood of the nodes, however, where a bundle is leaving the stele towards the leaf, the nature of the secondary tissues may change. Usually they become parenchymatous. This change affects the secondary xylem formed by the sector of cambium within the bundle itself: it may also affect neighbouring sectors of cambium. In addition, the interfascicular cambium that forms across the gap left by the trace may also produce parenchyma. Jeffrey (1917) referred to this tissue associated with a departing leaf trace as confronting and flanking parenchyma, and attached considerable importance to it in his theories on the evolution of the herbaceous stem.

As growth continues, the cambium near the departing leaf traces comes to produce normal secondary tissues. In deciduous trees the leaf traces are soon ruptured, but they may persist for some years in evergreen woody plants. In that event, the traces increase in length in step with the secondary growth of the stem. Once the leaf has fallen further extension ceases and the trace becomes ruptured. The cambium of the stem then becomes continuous, and subsequent secondary growth consists of normal xylem. All trace of nodal structure therefore is obliterated from xylem formed after the first few growth rings. Very rarely the specialized areas of cambium related to leaf-trace extension persist, so that in *Araucaria*, for example, leaf traces persist in the wood of the mature tree.

In certain species of *Dahlia* the cambium within some leaf traces may become inactive and take no part in the formation of the complete cambial ring. In its place cambial sectors form external to these leaf traces, which are thus left as complete bundles on the inner edge of the secondary xylem (Davis, 1961).

III. *The effect of gravity*

Frequently, characteristic differences can be observed between the cambium and its derivatives, as seen in erect stems and in lateral, more or less inclined, branches. These differences are part of the normal range of differentiation of the plant, but since they are closely related to the external stimulus of gravity, they will be considered more fully in a later chapter, when other environmental factors which can influence cambial structure and activity will be considered.

IV. *Patterns of differentiation in branch crotches*

The angle where branches meet is frequently marked by distinct folds or swellings. As the main branch increases in thickness, the base of the lateral branch becomes embedded in it. The cambium at the base of the lateral branch must therefore either become folded or undergo shortening (Jost, 1901). There are pronounced anomalies in the anatomy of the wood in this region.

V. *Morphological characteristics of the cambium in roots*

Only rarely have comparisons been made between the structure of the cambium in the roots and shoots of the same species, but some information can be derived from the few comparisons of the wood in roots and stems. Patel (1965) found that the rays in root-wood were larger both in cell number and cell size, and that in woods with uniseriate rays fewer of these occurred in the root. His results confirmed earlier morphological comparisons (De Bary, 1884). Differences of this kind must be greater in roots with specialized storage tissue, where the types of cells and their arrangement is very different from that in the aerial stems.

Because of its different origin (see Chapter 1), the cambium is smaller in diameter in a root than in the shoot of the same species. Consequently, in species with a storeyed cambium development of the storeys will begin nearer the centre of the axis in roots, and will proceed rapidly because of the great increase in circumference during the early stages of secondary growth. For this reason the blocks of initials in the cambium of roots will be larger than those in stems of similar size (Beijer, 1927; and see Chapter 5). For the same reason it is probable that the rapid increase in the size of cambial initials so characteristic of the core wood of stems with a non-storeyed cambium will be even more evident in roots, though no measurements of this effect appear to have been made.

The division of the root system of a tree into specialized regions or various types of root has often been observed. The subdivision of the root system is accompanied by modifications of the cambium. Wilson (1964) investigated the root system of *Acer rubrum* L., and distinguished: (*i*) a zone of taper in the main roots near the base of the trunk; (*ii*) long horizontal woody roots; and (*iii*) non-woody roots which are short-lived and in which there is little or no cambial activity. Anatomically the tapering roots near the base of the tree are more like stems than those roots which run through the soil at some distance from the tree. The taper is due to the asymmetrical and decreasing activity of the cambium and is accentuated by exposure to high winds (Jacobs, 1939). The horizontal woody roots of *Acer rubrum* are essentially cylindrical for considerable distances. The earlier growth rings of these roots are relatively concentric and wide, indicating a cambium which was at that time more or less uniformly active around its whole circumference. Towards the outside there is a change, usually abrupt, to much narrower growth rings, many of which are asymmetric or even discontinuous circumferentially. If these partial rings are traced back through the roots towards the trunk some will be found to disappear. That is to say, they are also discontinuous longitudinally. The sectors with narrowest rings are associated with branch roots, and Wilson suggests that the depression of cambial activity is due to the diversion of materials into the branch root.

A similar account of the distribution of radial growth, based on a study of species of the coniferous genus *Pinus*, has recently been given by Fayle (1968). He found that the growth layer usually tapers from the base of the tree towards the tip of the root and that this taper is generally rapid near the trunk. There may be local increases in radial thickness, but these display no consistent pattern. He frequently found the growth layer to be distributed around the roots eccentrically, especially near the base of the tree. Close to the trunk greater radial growth may occur on the upper side, and farther from the tree the growth may be greater on the lower side. Beyond this, eccentric growth displays no consistent pattern. Fayle compared the yearly pattern of ring width across the stem with that of the main root and found that they might show similar patterns. In subsidiary roots there was less or no correlation with the annual changes in increment in stem diameter.

In addition to differentiation between the cambium in stem and roots, and between various parts of the root system, Wilson (1964) also records differences in structure between sectors of the same growth ring. These differences, which are analogous to those between fascicular and interfascicular sectors of the stem cambium, are confined to the first or second growth rings. In *Acer rubrum* the xylem of the first one or two years' growth does not have vessels in the sectors opposite to the protoxylem. De Bary (1884) lists examples of roots in which rays develop opposite the protoxylem, and others in which the secondary xylem of the root is uniform around the whole circumference. Broad rays opposite the protoxylem are conspicuous in the young roots of *Griselinia lucida* Forst.f. In older roots the sectors of the cambium which have produced these rays no longer give rise to parenchyma, and the ray becomes replaced by normal xylem (Mason, R. (1937), and Fig. 6.4). *Griselinia lucida* is usually an epiphytic tree, and the roots described by Mason

Fig. 6.4 Griselinia lucida. T.S. mature root. Sectors of the cambium opposite the protoxylem groups produce distinctive derivatives for several years. Eventually the secondary xylem becomes uniform throughout. (×1.) (Photograph by courtesy of Dr Ruth Mason, Christchurch, N.Z.)

were descending aerial roots. It may be significant that the structure of these aerial roots, while in the young state, is similar to that of the stems of some lianes. The body of the xylem is divided into radiating bands by wide and deep plates of parenchyma. It may well be that the stresses to which these roots are subjected are similar to those met by the stems of woody climbers.

VI. *The cambium in haustoria*

An example of a greatly modified cambium found in an organ that does not fit readily into the normal categories of organs is given by Fineran (1963) in his description of the haustoria of *Exocarpus*. The haustorium of this woody root parasite is not homologous with a branch root, but is an organ arising exogenously on the parent root (Fineran, 1965). A cambium forms within the haustorium which is continuous with the cambium of the root bearing it. The inner derivatives of this cambium differentiate either as parenchyma or as tracheary elements. As might be expected, there is no secondary phloem, since no phloem elements are present in the primary haustorium. Instead, the outer derivatives of the cambium mature as secondary cortical cells. The suckers which arise from the haustorium and which penetrate to the vascular tissue of the host root also increase by regular, cambium-like, divisions. In this way they keep pace with the secondary increase of the host root. This cambium-like meristem of the suckers was shown, in a few instances, to be continuous with the cambium within the haustorium itself.

REFERENCES

BALFOUR, E. E. (1958). The development of the vascular systems of *Macropiper excelsum* Forst. II. The mature stem. *Phytomorphol.* **8,** 224–33.

BARGHOORN, E. S. (1964). Evolution of cambium in geological time. In ZIMMERMANN, M. H., *The Formation of Wood in Forest Trees*, pp. 3–17, Academic Press, New York.

BEIJER, J. J. (1927). Die Vermehrung der radialen Reihen im Cambium. *Rec. trav. Bot. Neerl.* **24,** 631–786.

CATESSON, A. M. (1964). Origine, fonctionnement et variations cytologiques saisonnières du cambium de l'*Acer pseudoplanatus* L. (Acéracées). *Ann. Sci. nat. (Bot.)* (12e sér.) **5,** 229–498.

DAVIS, E. L. (1961). Medullary bundles in the genus *Dahlia* and their possible origin. *Am. J. Botany* **48,** 108–13.

DE BARY, A. (1884). *Comparative Anatomy of the Vegetative Organs of Phanerogams and Ferns*, Oxford University Press, Oxford.

FAYLE, D. C. F. (1968). Radial Growth in Tree Roots. *Faculty of For. Tech. Rept. No. 9*, Univ. Toronto, pp. 1–183.

FINERAN, B. A. (1963). Studies on the root parasitism of *Exocarpus bidwillii* Hook.f. – IV. Structure of the mature haustorium. *Phytomorphol.* **13,** 249–67.

FINERAN, B. A. (1965). Studies on the root parasitism of *Exocarpus bidwillii* Hook.f. – V. Early development of the haustorium. *Phytomophol.* **15,** 10–25.

FREY-WYSSLING, A. (ed.) (1964). *Multilingual Glossary of Terms used in Wood Anatomy,* Konkordia Winterthur.

JACOBS, M. R. (1939). Study of the effect of sway on trees. *Commonwealth Forestry Bureau Australia, Bull. No. 26* (17 pp.).

JEFFREY, E. C. (1917). *The Anatomy of Woody Plants,* Chicago University Press, Chicago.

JOST, L. (1901). Über einige Eigenthümlichkeiten des Cambiums der Bäume. *Botanische Zeitung, Berlin* **59,** 1–24.

MASON, R. (1937). Some observations on *Griselinia lucida.* M.Sc. thesis, University of Auckland, New Zealand.

PATEL, R. N. (1965). A comparison of the anatomy of the secondary xylem in roots and stems. *Holzforschung* **19,** 72–9.

WILSON, B. F. (1964). Structure and growth of woody roots of *Acer rubrum* L. Harvard Forest paper No. 11, pp. 1–14, Petersham, Mass.

7
Anomalous cambia

It is perhaps not always fully appreciated that anomalous means of secondary thickening are widely dispersed among dicotyledons, nor that the anomalies are of several diverse types. These interesting variations have been reviewed by several authors, so that no detailed description of them need be repeated here, nor need the complete range of the species showing each type of anomaly be listed. This information is contained in the following: de Bary (1884), Schenck (1893). Pfeiffer (1926), Metcalfe and Chalk (1950), Boureau (1957), Obaton (1960), Philipson and Ward (1965).

A number of classifications of cambial anomalies are proposed in the works just cited. A simple subdivision into two main classes, each further subdivided, will be adopted here. Cambia which occur in the normal position, but whose activity is atypical in some way, form the first of these classes. The second includes cambia which occur in unusual positions in the stem or root.

I. Cambia in the normal position, but atypical

1. *Asymmetric activity*

Secondary growth may take place at unequal rates on different parts of the circumference of the cambium. This is not at all unusual in woody plants, resulting in fluted or excentric stems, Newman (1956). Glock and Agerter (1962) give interesting information about growth increments of limited circumferential and vertical extent in stems, and Wilson (1964) in roots. While it is often difficult to determine the cause of this uneven growth in girth, sometimes it is clearly related to the asymmetrical arrangement of foliage or branches. In fastigiate trees, for example, all the foliage and subsidiary twigs are often confined to one side of the branches, and the growth of the cambium is much stronger on that side. Greatly exaggerated asymmetrical cambial activity may be a genetically determined character, as in trees with buttress roots. Genetical asymmetry is pronounced in some lianes which develop stems of unusual and characteristic shapes. If growth is retarded over two opposite arcs of cambium and accentuated elsewhere a flattened ribbon-like stem will result (Basson and Bierhorst, 1967). Or if the arcs of retardation are small the stem will come to have a cross-section shaped like a figure 8. If several segments of retardation and accentuation alternate around the cambium the stem will become lobed with a star-shaped cross-section.

2. *Cessation of xylem formation*

In lianes belonging to several families, but notably the Bignoniaceae, Hippocrateaceae, Icacinaceae, Apocynaceae, Acanthaceae (Mullenders, 1947), and Passifloraceae (Ayensu and Stern, 1964), xylem formation ceases, or is greatly retarded, over short arcs of the cambium. The external surface of the stem remains approximately cylindrical, but the surface of the xylem becomes deeply furrowed. As the stem increases in diameter the furrows may increase in width by the cessation of xylem production over adjacent arcs of cambium. In addition, new furrows may be initiated by the cessation of xylem production in other positions. In this way the xylem becomes star-like in transverse section (Fig. 7.1). The extreme development of this anomaly leads to stems with irregularly dissected (dispersed) steles. The furrows between the arms of xylem become filled with external derivatives of the cambium (phloem). The cambium, which remains at the external surface of the xylem at the base of the furrows, must produce sufficient external derivatives to fill these furrows. It will be appreciated that, since the phloem which fills the furrows is formed at their base, there must be movement of the phloem relative to the xylem which forms the furrows. As de Bary (1884) says 'a continuous displacement goes on between . . . the lateral faces of the bast-plate and the adjoining ones of the woody projections; the two faces are not grown together one with another; in transverse sections, even of fresh internodes, a slit-like space often appears between the two' (Fig. 7.2).

3. *Modified interfascicular cambium*

In *Aristolochia* the rays are prominent. At the same time as the original primary bundles increase in size by means of the fascicular cambium the rays are also extended by additional thin-walled tissue. However, no clearly defined interfascicular cambium can be detected. As the diameter of the stem increases, the wedges of secondary xylem become dissected by the modification of arcs of cambium. These cease to produce secondary xylem and phloem, and in their place produce rays of parenchyma, which resemble the original rays in all respects except their more limited vertical extent. These modified

Fig. 7.1 A, Deeply furrowed secondary xylem of *Passiflora* sp., and B, dissected stele of *Passiflora multiflora* L. Photographs by courtesy of Dr W. L. Stern. (A × 2·3; B × 2.)

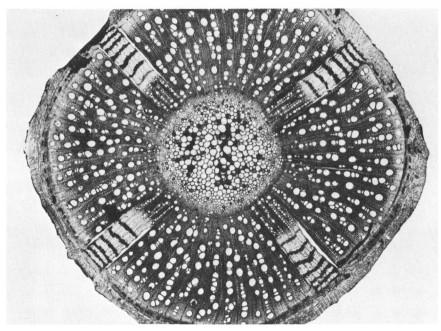

Fig. 7.2 Stem of *Pyrostegia venusta* (Ker-Gawl.) Miers in transverse section. Four sectors of the cambium form little xylem but considerable phloem. Growth movements between adjacent sectors result in prominent radial longitudinal splits. (× 10.)

arcs of cambium also become ill-defined, as in the interfascicular regions. In *Tetrapatheae* and *Clematis* the cambium which produces wide parenchymatous rays is similarly ill defined, but in *Clematis* the interfascicular cambium in some species may eventually produce normal secondary wood. This wood is referred to as intermediate bundles by de Bary (1884). Where these occur, the cambium becomes clearly defined and normal in appearance. The related genus *Atragene* has a complete ring of normal cambium from the beginning of secondary thickening, and it is normal for the interfascicular sectors to be woody ¡ather than parenchymatous. It is possible that *Atragene*, being less strongly climbing in its growth, represents the normal type of dicotyledonous stem from which the modified stem of *Clematis* has been derived.

The three types of stem just described are characteristic of lianes, and each achieves a similar result by different means. The stem of a liane faces different mechanical problems from that of an erect self-supporting plant. It consists of a more or less pliant axis attached at points frequently far apart. It must therefore be capable of torsion movements without damage to the woody conducting elements. This is achieved by the subdivision of the xylem cylinder, either by lobing as in (1) *asymmetric activity*, by dissection as in (2) *cessation of xylem formation*, or by the retention and accentuation of the separate bundles present in the primary condition, as in (3) *modified interfascicular cambium.*

4. *Included phloem:* Combretum *type*

In some plants strands of phloem are embedded within the mass of secondary wood. This intraxylary phloem may arise in several distinct ways. Two types of intraxylary phloem are described in this and the following paragraph. Others will be considered at the end of this chapter. In some species of *Combretum* the internal derivatives of small arcs of cambium may, for a short time, differentiate as phloem instead of xylem. These small phloem strands become embedded in secondary xylem as soon as the normal activity of the cambium is resumed (Fig. 7.3). Similar abnormalities are recorded for *Salvadora* (Scott and Brebner, 1889), *Leptadenia* (Balwant Singh, 1943), *Thunbergia* (Mullenders, 1947), and *Lebrunia* (Duchaigne, 1951).

5. *Included phloem:* Strychnos *type*

In species of *Strychnos*, xylem production may be retarded over short arcs of cambium so that phloem becomes slightly sunken in grooves on the surface of the xylem cylinder. The circular outline of the cambium becomes restored by new arcs of cambial initials forming outside the sunken parts of the phloem (Fig. 7.3). This anomaly is very closely related to the formation of successive cambia in, for example, *Avicennia*, as described later in this chapter.

II. Cambia in unusual positions

1. *Medullary cambia*

Of the many plants with phloem strands in the pith, few develop a second cambium in association with this additional phloem. When secondary phloem is added to these internal phloem strands it is formed by a cambium initiated between the protoxylem and the primary internal phloem. Such medullary cambia function in the same way as normal cambia, forming phloem and xylem. However, the relative positions of these two tissues is reversed. Secondary xylem is added between the protoxylem and the internal cambium; secondary phloem is added between the internal cambium and the primary internal phloem. The result is that inverted bundles, or continuous cylinders, are formed (Fig. 7.3).

Secondary thickening occurs within complete medullary bundles in a few plants; for example, this was recorded by Balfour (1958) for *Macropiper excelsum* (Forst.f.) Miq. Medullary cambia may also arise independently of either phloem strands or complete bundles. For example, Scott and Brebner (1891) describe the initiation of a cambium in the pith of *Acantholimon*. The secondary vascular tissues which it produces are unrelated to any primary tissues.

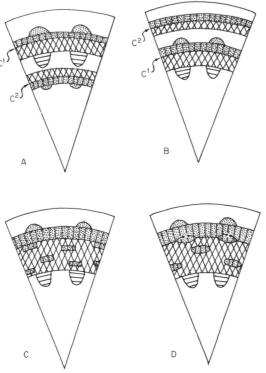

Fig. 7.3 The principal variants of the cambium (for explanation of symbols see Fig. 7.5, p. 102). C¹, first cambium; C², second cambium. A, additional cambium associated with medullary phloem strands; B, additional cambium external to the primary stele; C, cambium producing intra-xylary phloem, *Combretum* type; D, cambium producing intra-xylary phloem, *Strychnos* type. (From Philipson and Ward, 1965.)

2. *Cortical cambia*

Slight secondary thickening may occur in cortical bundles, as for example, in the *Calycanthaceae*. Most cambia external to the normal positions, however, arise without relation to pre-existing primary vascular tissue. In *Avicennia* a succession of cambia replace one another, each functioning for a short time. The woody cylinder of the branches, therefore, consists of concentric alternating circles of xylem and phloem (Fig. 7.4). The first cambium arises in the normal position between the primary xylem and phloem. The second

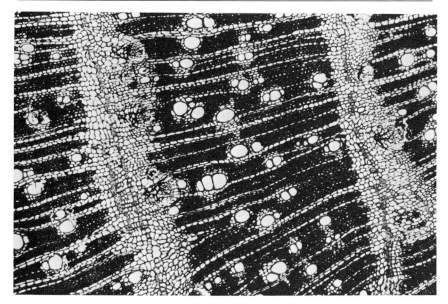

Fig. 7.4 Avicennia. Transverse section of mature stem: broad bands of fibres and vessels with conspicuous rays alternate with bands of parenchyma in which there are phloem groups and narrow bands of sclereids. (From Studholme and Philipson, 1966.) (× 40.)

cambium arises immediately outside the pericyclic fibres (Studholme and Philipson, 1966). As the first internal derivatives of this second, and all subsequent cambia, mature as parenchyma, it soon appears as though the cambium had arisen in the outer cortex, as in fact de Bary (1844) believed. Subsequent cambia in *Avicennia* continue to arise external to the phloem. Obaton (1960) describes a different origin for the first additional cambium in *Salacia.* Here the cambium lies internal to the ring of pericyclic fibres, and so forms within the phloem. Other examples of plants with successive external cambia are *Cocculus* (Menispermaceae), *Phytocrene* (Icacinaceae), *Acantholimon* (Plumbaginaceae), *Bauhinia* (Leguminosae), and *Boscia* (Capparidaceae). In the last example Adamson (1936) describes a complex arrangement of external cambia which produce vascular tissue in an anomalous fashion. External cambia have been recorded in several other genera, but in the absence of developmental studies the precise type of anomaly present cannot always be decided.

Very anomalous stems are found in Sapindaceous lianes. Circular cambia may arise in the cortex so that several additional steles develop. Obaton (1960) has shown how, in *Paullinia pinnata* L., these steles connect with the central ring of bundles at the nodes, and how the cambium of these external steles is continuous with the normal cambium of the stem. In some of these lianes no stele is present in the normal central position. An investigation of the development of such a stem would be of great interest. It might be found to have a similar origin to the poly-stelic stems of, for example, some Umbelliferae, where a cambium surrounds the individual primary bundles.

The multiplication of active cambia in the stems of some climbing members of the Sapindaceae, Malpighiaceae, and Acanthaceae may be so great that stems with irregularly dispersed steles are formed. These stems are superficially very like the extreme examples of Bignoniaceous, Apocynaceous, or Passifloraceous lianes already referred to. Other excessively anomalous stems occur in species of *Bauhinia* (Handa, 1937; 1938) and *Iodes* (Obaton, 1960). In the root of the sweet potato, *Ipomoea batatas* (L.) Lam., vessels or groups of vessels become surrounded by cambia which produce phloem and xylem rich in storage parenchyma. Stems or roots in which several distinct steles develop may split into a number of separate

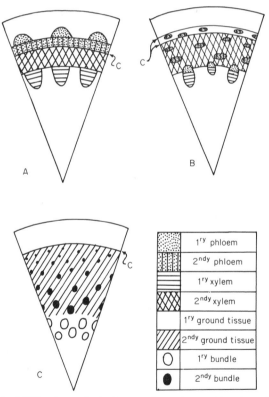

1^{ry} phloem	
2^{ndy} phloem	
1^{ry} xylem	
2^{ndy} xylem	
1^{ry} ground tissue	
2^{ndy} ground tissue	
1^{ry} bundle	
2^{ndy} bundle	

Fig. 7.5 The three principal types of vascular cambium in angiosperms: A, the normal type of cambium; B, the type found in Nyctaginaceae, Chenopodiaceae, Amarantaceae, and a few other families; C, the monocotyledonous type. Symbol C cambium.

strands, each with its own means of secondary thickening and with its own corky outer layer (Moss and Gorham, 1953).

Opinions differ as to whether a type of anomalous cambium found in the families Chenopodiaceae, Amarantaceae, Nyctaginaceae, and Phytolaccaceae, together with a few species of Compositae and Stylidiaceae, should be regarded as an example of successive cambia, or whether it represents a lateral meristem with a distinctive mode of activity. The occurrence of a recognizable anomalous type within all the members which show secondary growth of certain taxonomic groups is of considerable interest from the point of view of comparative morphology. This type of cambium, therefore, has been studied repeatedly, and since various interpretations have been placed upon it, it will be considered here in greater detail than the anomalies in the preceding pages.

The type of thickening meristem found in these families was called a uni-directional cambium by Philipson and Ward (1965) because all the mature vascular tissue derived from it, both phloem and xylem, lies to one side (Fig. 7.5). In contrast, the normal cambium was called bi-directional. While this distinction is valid, the application of names over-emphasizes the importance of this anomaly, and so we do not retain them here.

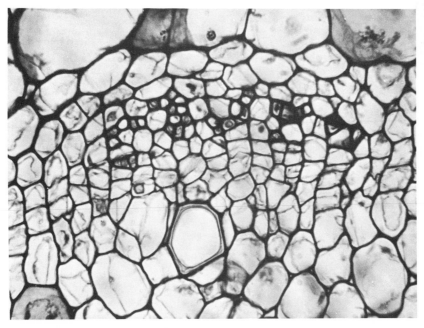

Fig. 7.6 Heimerliodendron. T.S. cambial zone, with a group of several phloem elements and a single prominent vessel. (× 325.) (From Studholme and Philipson, 1966.)

The cambial zone in these families consists of a meristem within which an inner region of more active division can be detected. The outer cells of the zone, however, retain the capacity to divide, and indeed the balance of activity may shift to these outer derivatives. Therefore, instead of a permanent initial layer, the active cambium constantly moves into its own outer derivatives. Phloem formation occurs by differentiation of some outer cells and has been described in detail by Esau and Cheadle (1969). Differentiating phloem may be recognized in transverse section by the formation of clusters of small cells (Fig. 7.6). These arise by longitudinal division of cells derived in radial

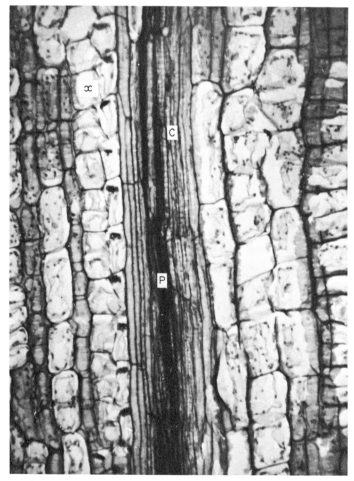

Fig. 7.7 Heimerliodendron. R.L.S. cambial zone. The difference in length between the cambial and the cortical cells is pronounced. C, cambium; P, phloem strands in centre of the cambial zone; X, cortex. (×170.) (From Studholme and Philipson, 1966.)

sequence from the active cambial region. The cells which divide and form the phloem strands are never the outermost of the meristematic zone, since these are involved in the continuation of secondary thickening. This cannot be detected by the examination of transverse sections, because in them the outer cells of the thickening meristem can be confused with inner cells of the cortex. By the examination of radial longitudinal sections Studholme and Philipson (1966) were able to distinguish between cambial and cortical cells (Fig. 7.7). If the cambium moved outwards by invading cortical cells the meristem would come to have cells of the same length as those of the cortex. It is clear from the radial longitudinal sections that the cambium moves outwards by divisions within its own cells. The phloem bundles initiated in the way just described soon come to lie to the inner side of the meristem. This is due to the maturation of the inner cells of the cambial zone. At the same time as these inner cells are lost from the cambium they are replaced by divisions in the peripheral cells of the cambium. The identity of the meristem is thereby preserved, although it moves outwards continuously as the secondary tissues increase in diameter.

The inner derivatives differentiate as secondary xylem. The xylem consists of a ground tissue of fibres or parenchyma in which groups of vessels occur. These groups of vessels are usually associated with the groups of phloem whose formation has been described already (Fig. 7.8). These associations of

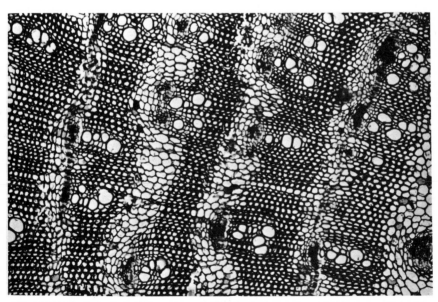

Fig. 7.8 Heimerliodendron. T.S. mature stem. Bands of fibres and vessels alternate with bands of parenchyma and phloem. (× 65.) (From Studholme and Philipson, 1966.)

phloem and vessels show a superficial resemblance to the secondary bundles of a woody monocotyledon, though their origin is quite different, and the phloem and xylem behave independently in their longitudinal course (Fahn and Shchori, 1968).

These cambia typically produce no permanent external derivatives, because division is active in its outermost cells. However, in *Salicornia*, for example, external cells cease to divide and become differentiated as parenchyma. The tissue formed by these cells may be referred to as secondary cortex. No vascular tissue persists external to these anomalous cambia.

Not only is the action of the cambium distinctive but the manner of its initiation in the young stem may also be different from that of a normal cambium. Balfour (1965) has described three ways in which these cambia may arise. In *Chenopodium* a cambium arises in the normal position between the phloem and xylem of the primary bundles, and later extends across the medullary rays to form a complete ring. This cambium remains active for a very limited time: its cells differentiate and the function of cell-initiation passes to cells lying outside the primary phloem. The cells appear to be a residuum of the original procambial ring whose meristematic activity is resumed. In *Iresine* the primary bundles do not become linked by interfascicular cambium. The first complete cambial ring forms in the residual procambial ring in contact with the outer faces of the phloem. In *Bougainvillea* and *Heimerliodendron* the cambium arises without any relation to the primary vascular system or to the procambial ring. Semi-mature cortical cells well below the apical meristem and well separated from the primary bundles resume active division, thus forming a thickening zone which encircles the stem (Fig. 7.9). Adamson (1934; 1937) has described a similar origin for the cambium of some South African shrubby members of the Compositae, and Mullenders (1947) for the cambium of a species of *Stylidium*.

A number of interpretations of this type of cambium have been put forward. First, the cambium was thought to be similar to that found in some monocotyledons (see Chapter 8, and Fig. 7.5), that is to say, a single cambium persists and gives rise internally to conjunctive tissue and vascular bundles. This interpretation was held by Sanio (1863) and de Bary (1884), but has found few adherents in the present century. It is true that in some stems the gross appearance of the tissues is similar to that of monocotyledons of the *Cordyline* type. Associated groups of phloem and xylem resembling bundles embedded in conjunctive tissue are left behind by a cambium that moves continuously outwards. However, detailed examination shows that the operation of the cambium is distinct from that in monocotyledons. In that group the complete bundles are formed as units derived by the repeated and randomly orientated division of mother-cells, which in turn are inner derivatives of the cambium. In these dicotyledonous anomalous cambia the xylem

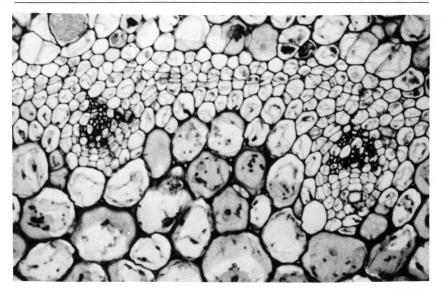

Fig. 7.9 Heimerliodendron. T.S. cambial zone soon after its initiation, with two groups of vascular tissue. (× 150.) (From Studholme and Philipson, 1966.)

forms by the direct maturation of internal derivatives, and the phloem arises by the division and maturation of outer derivatives of the active cambial cells. The description of *Stylidium adnatum* R. Br. given by Mullenders (1947) speaks of a cambium which produces both xylem and phloem to the inner side. However, he emphasizes that the xylem and phloem are not associated as bundles, as in the secondary tissues of monocotyledons. It is clear from his account that the cambium of that plant closely resembles that of *Chenopodium*.

A second interpretation regards the cambium as functioning normally, producing xylem to the inside and phloem to the outside, but that it operates only for a limited time, being replaced by new cambia, or arcs of cambium, which form successively farther from the centre of the stem. On this view the anomaly would be similar to that already described for *Cocculus* and *Strychnos*. This interpretation was put forward as early as 1885 by Morot, and was supported by de Fraine (1913) and Artschwager (1920), among many others. It has been adopted in standard works such as Pfeiffer (1926), Metcalfe and Chalk (1950), and Esau (1965). No clearer statement of this view has been given than that of Scott and Brebner (1889): 'The cambium constantly gives rise on its inner side to the wood of the bundles, on its outer side to their bast. The generative zone always passes between *xylem* and *phloem*, but only as a temporary activity. In order that other more external bundles may be formed, it is necessary for new generative zones to be produced outside the first one. These new cambial zones may either be entirely distinct from their predecessors, or, as is more frequent, may have more or less numerous points of contact with them. In the latter case the original generative zone may maintain its activity in the intervals between the bundles, but these interfascicular cambial arcs become connected by bridges of meristem passing outside the bast, and destined to replace the generative arcs interposed between phloem and xylem.' A recent detailed argument in favour of this interpretation has been set out by Esau and Cheadle (1969).

A third interpretation places importance on differences which exist between the cambial anomalies found in the Chenopodiaceae and the other associated families and those anomalies known as successive cortical cambia. While the interpretation of Scott and Brebner just quoted would regard these as variants on a single theme, Studholme and Philipson (1966) emphasized their differences. They compared the cambium of *Heimerliodendron* (Nyctaginaceae) with that of *Avicennia* (Verbenaceae). They were impressed by the fact that the site of maximum division, in *Heimerliodendron*, passes to other cells of the cambial zone, so that the cambium moves constantly outwards, leaving behind it a cylinder of secondary vascular tissue. In *Avicennia*, on the other hand, the cambium functions normally for some time before being replaced by a new cambium which arises in cortical cells. Later cambia, however, arise in the outer derivatives of previous ones. These authors further considered that phloem formation also deviated from the normal, but the later work of Esau and Cheadle (1969) shows this to be unfounded. However, the important feature that the outermost cells of the cambial zone remain undifferentiated, enabling cambial activity to progress outwards, remains valid.

While they preferred to recognize the distinctness of the anomaly, and indeed gave the type a name, Studholme and Philipson discussed the possible relation between it and the condition in *Avicennia*. It is not impossible to imagine how one might have been derived from the other, and both from the typical cambium. If the external derivatives of a normal cambium were to retain the power of division, and the initials were to lose it, the main feature of the anomaly would have been achieved. In fact, a situation approaching this has been reported in *Salacia cordifolia* Hook. f. (Obaton, 1960), where the successive cambia arise by the outer phloem cells resuming meristematic activity.

The differences between the second and third interpretations may be slight, but may yet be of interest from the viewpoint of comparative morphology. Unfortunately any consideration of the phylogenetic relationship of these anomalies can only be speculative. Adamson was undoubtedly correct when he claimed that this feature has arisen independently in three or more lines within the Compositae. His explanation, that herbs which had lost the power of secondary thickening later re-evolved as woody plants, seems plausible. However, it is remarkable that in doing so they did not re-assume the normal cambium found throughout that family. The situation in the Chenopodiaceae, Nyctaginaceae, and Amarantaceae is rather different, because in these families anomalous cambia occur in all species with secondary thickening. Possibly some herbaceous form, ancestral to one or all of these families, lost a normal cambium and on resuming secondary growth acquired one of distinctive type. An alternative hypothesis, that this group of families diverged from the dicotyledonous stock before the acquisition of normal secondary thickening, has little to recommend it.

REFERENCES

ADAMSON, R. S. (1934). Anomalous secondary thickening in Compositae. *Ann. Botany* **48**, 505–14.

— (1936). Note on the stem structure of *Boscia rehmanniana* Post. *Trans. Ry. Soc. S.Afr.* **23**, 297–301.

— (1937). Anomalous secondary thickening in *Osteospermum*. *Trans. Ry. Soc. S.Afr.* **24**, 303–12.

ARTSCHWAGER, E. F. (1920). On the anatomy of *Chenopodium album* L. *Am. J. Botany* **7**, 252–60.

AYENSU, E. S. and STERN, W. L. (1964). Systematic anatomy and ontogeny of stem of Passifloraceae. *Contr. U.S. Nat. Herb.* **34**, 48–73.

BALFOUR, E. E. (1958). The development of the vascular systems of *Macropiper excelsum* Forst. II. The mature stem. *Phytomorphol.* **8**, 224–33.

— (1965). Anomalous secondary thickening in Chenopodiaceae, Nyctaginaceae and Amarantaceae. *Phytomorphol.* **15**, 111–22.

BALWANT-SINGH (1943). The origin and distribution of inter- and intraxylary phloem in *Leptadenia. Proc. Indian Acad. Sci.* **18**, B, 14–19.

BASSON, P. W. and BIERHORST, D. W. (1967). An analysis of differential lateral growth in the stem of *Bauhinia surinamensis. Bull. Torrey Bot. Club* **94**, 404–11.

BOUREAU, E. (1957). *Anat. vég.* **3**, U.P. de France, Paris.

DE BARY, A. (1884). *Comparative Anatomy of the Vegetative Organs of Phanerogams and Ferns*, Oxford University Press, Oxford.

DE FRAINE, E. (1913). The anatomy of the genus *Salicornia. J. Linn. Soc.* **41**, 317–48.

DUCHAIGNE, A. (1951). L'ontogénie du phloème intraxylémien dans la tige du *Lebrunia bushaie* Staner (Gutifferes). *C.R. Acad. Sci., Paris* **232**, 646–8.

ESAU, K. (1965). *Plant Anatomy* (2nd ed.), Wiley, New York.

— and CHEADLE, V. I. (1955). Significance of cell divisions in differentiating secondary phloem. *Acta Bot. Neerl.* **4**, 348–57.

— and — (1969). Secondary growth in *Bougainvillea. Ann. Botany* **33**, 807–19.

FAHN, A. and SHCHORI, Y. (1968). The organization of the secondary conducting tissue in some species of the Chenopodiaceae. *Phytomorphol.* **17,** 147–54.

GLOCK, W. S. and AGERTER, S. R. (1962). Rainfall and tree growth. In KOZLOWSKI, T. T., *Tree Growth,* Ronald Press, New York.

HANDA, T. (1937). Anomalous secondary thickening in *Bauhinia japonica* Maxim. *Jap. J. Botany* **9,** 37–53.

— (1938). Anomalous secondary growth in the axis of *Bauhinia championi* Benth. *Jap. J. Botany* **9,** 303–11.

METCALFE, C. R. and CHALK, L. (1950). *Anatomy of the Dicotyledons,* Oxford University Press, Oxford.

MOROT, L. (1885). Recherches sur la péricycle ou couche péripherique du cylindre central chez les Phanérogames. *Ann. Sci. nat. (Bot.) ser. 6,* **20,** 217–309.

MOSS, E. H. and GORHAM, A. L. (1953). Interxylary cork and fission of stems and roots. *Phytomorphol.* **3,** 285–94.

MULLENDERS, W. (1947). L'origine du phloème interxylemien chez *Stylidium* et *Thunbergia. La Cellule* **51,** 7–48.

NEWMAN, I. V. (1956). Pattern in meristems of vascular plants. 1. Cell partition in living apices and in the cambial zone in relation to the concepts of initial cells and apical cells. *Phytomorphol.* **6,** 1–19.

OBATON, M. (1960). Les lianes ligneuses à structure anomale des forêts denses d'Afrique occidentale. *Ann. Sci. nat. (Bot.) ser. 12,* **1,** 1–220.

PFEIFFER, H. (1926). Das abnorme Dickenwachstum. In LINSBAUER, K., *Handbuch der Pflanzenanatomie,* Bd. IX, Bornträger, Berlin.

PHILIPSON, W. R. and WARD, J. M. (1965). The ontogeny of the vascular cambium in the stem of seed plants. *Biol. Rev.* **40,** 534–79.

SANIO, K. (1863). Vergleichende Untersuchungen über die Zusammensetzung des Holzkörpers IV. *Bot. Zeit.* **21,** 401–12.

SCHENCK, H. (1893). *Beiträge zur Biologie und Anatomie der Lianen, II,* Jena.

SCOTT, D. H. and BREBNER, G. (1889). On the anatomy and histogeny of *Strychnos. Ann. Botany* **3,** 275–304.

STUDHOLME, W. P. and PHILIPSON, W. R. (1966). A comparison of the cambium in two woods with included phloem: *Heimerliodendron brunonianum* (Endl.) Skottsb. and *Avicennia resinifera* Forst.f. *N.Z. J. Botany* **4,** 355–65.

WILSON, B. F. (1964). Structure and growth of woody roots of *Acer rubrum* L. Harvard Forest paper No. 11, pp. 1–14, Petersham, Mass.

8

The thickening
of stems in
monocotyledons

Many botanists have considered that traces of cambial activity can be found in the vascular bundles of numerous monocotyledons. Dr Agnes Arber, in particular, published a series of papers summarizing previous records of an intra-fascicular cambium in a wide range of monocotyledons and added many additional observations of her own (Arber, 1917; 1918; 1919; 1925). In these investigations a cambium was considered to be present if more or less regular tangential divisions could be seen in meristematic cells between the xylem and phloem, as seen in transverse sections, or if the elements of the xylem and phloem were arranged in radial files.

Whether these characters can be accepted as conclusive evidence for the presence of a cambium depends on the definition of cambium used. The criteria for the recognition of a cambium have been discussed in relation to the dicotyledons in Chapter 1. The files of cells sometimes seen in monocotyledon vascular bundles are often most striking in young stems still capable of growth in length, and are then best interpreted as due to regularly orientated divisions of procambial cells. When the stem persists from one growing season to the next and the serial production of cells is resumed after a resting period (*Veratrum, Gloriosa*) growth in length may have ceased. But even in such plants there is no evidence of the differentiation into fusiform and ray initials characteristic of dicotyledonous cambia.

The evidence for the existence of cambial growth in the vascular bundles of monocotyledons is therefore not conclusive. In any event, at best it can be no more than slight and ephemeral. In no case is it known to extend beyond the bundle as interfascicular cambium, nor does it contribute significantly to the increase in diameter of the stem. The regular orientation of the divisions is of interest in so far as it indicates similar ontogeny of the vascular bundles of monocotyledons and dicotyledons, but the complete absence of further development of a normal cambium points to a fundamental distinction in this respect between the two great subdivisions of the angiosperms.

This complete, or virtual, absence of a vascular cambium of the normal type from all monocotyledons, and the restriction of any method of increasing the thickness of stems and roots to very few species, has had a profound effect on habit in this large group of plants (Holttum, 1955). A root system which lacks a cambium will soon prove inadequate unless it is supplemented by adventitious roots. As these adventitious roots arise from the nodes, it is an advantage for the shoot, if erect, to have short basal internodes. This crowding of the lower nodes also allows the stem diameter to attain mature size near to ground level (this initial increase in stem thickness is discussed below), and the crowded roots also support the otherwise most unstable tapered base of such monocotyledonous stems as palms and maize. Once the mature stem-thickness has been attained, the stem is committed to future growth with mechanical support which cannot be increased and a constant vascular supply. The growth of such a shoot may be considerable, but it is severely limited in comparison with a dicotyledonous tree. Most monocotyledons, however, prolong the life of the individual plant by the production of basal side shoots. Holttum derives the many and varied habits to be found among monocotyledons from such a basic growth-form.

In spite of the absence of a normal vascular cambium, the stems of mono-cotyledons may increase in thickness in three distinct ways. These are: (*i*) by cell division during the phase of primary growth; (*ii*) by diffuse secondary thickening; and (*iii*) by the action of a type of secondary thickening peculiar to monocotyledons. These methods will be discussed separately.

Stem thickening during primary growth

Examples of monocotyledons with thick stems are the fleshy rhizomes of the Iridaceae or the Zingerberaceae, the bulbous storage organs of the Amaryllidaceae or Liliaceae, or the woody columnar stems of many palms. All these stems become thick quite close behind the growing point (Helm, 1936; Chouard, 1936; Eckardt, 1941). The apical meristem itself is not abnormally large, but immediately behind it cell division is remarkably active and results in the rapid increase in diameter of the stem. This increase takes place before primary extension growth has ceased. The activity must there-fore be regarded as part of the primary growth of the shoot. Further increase in diameter may take place below this meristemic region, but this is due mainly to the increase in size of cells already formed close to the apex, though further divisions occur as the tissues differentiate. Stems may therefore taper slightly towards their tips, but this is not due to secondary growth. Primary thickening of the stem becomes active early in the life of the epicotyl, but it can build up only gradually from the small diameter present in the embryo axis. Internodes tend to be short at the base of stems, with the result that the final stem size is attained near to the base of the plant.

The meristem, which provides the abrupt dilation of the originally narrow apex to a wide crown bearing congested leaf-primordia, lies below the youngest leaf bases (Fig. 8.1A). That it is massive can be appreciated by an examination of the stem apex of *Veratrum* illustrated by Clowes (1961, Plate 19, a). Within this meristematic region there may be distinguished localized mitotic activity which results in the formation of procambial strands running more or less horizontally and parallel to the surface of the apex (Fig. 8.1B). This belt of procambium formation is the meristematic cap described by Zimmermann and Tomlinson (1967), who clarify its rôle in the development of the bundle system of the stem of the palm *Rhapis*. The bulk of the meristem responsible for stem thickening lies below this cap, although much of it has been derived from the ground tissue which lies between the procambial strands of the cap (Figs. 8.1, A and B). The cap is distended by the cell increase below it which causes the enlargement of the stem in length and breadth. Because this meristematic cap can be seen readily in sections, authors attributed to it the principal rôle in stem building and described it as the primary thickening meristem. Zimmermann and Tomlinson (1968) recommended that use of the term primary thickening meristem be discontinued because its application has been restricted in this way. The cap, then, is to be regarded as a specialized part of a massive thickening region. It functions in a similar manner to the cambium of some monocotyledons (see later section of this chapter): in both, procambial strands are formed running through ground tissue in a direction parallel to the stem surface. Indeed, the cap and the cambium, when present, have been found to be continuous. This is not to say that the cambium is a downward prolongation of the meristem responsible for stem thickening in the primary phase of growth (as is often said) but only with this part of it. The means of thickening in the primary state contains within itself the origins of the secondary thickening meristem. The leaf primordia are partly vascularized before increase in stem diameter occurs, so that the main leaf traces become greatly distorted by the subsequent growth of the stem, but bundles which are added late in the development of the leaf are less distorted or run quite direct (Chouard, 1936).

Diffuse secondary thickening

The bulk of the trunk of palm trees is primarily due to the action of a thickening meristem of the kind just described. Yet, in spite of the absence of a cambium of any kind, many palm stems are capable of further increase in diameter. This swelling may be confined to the upper part of the stem, just below the crown of leaves, but in some palms, particularly in *Roystonea*, it

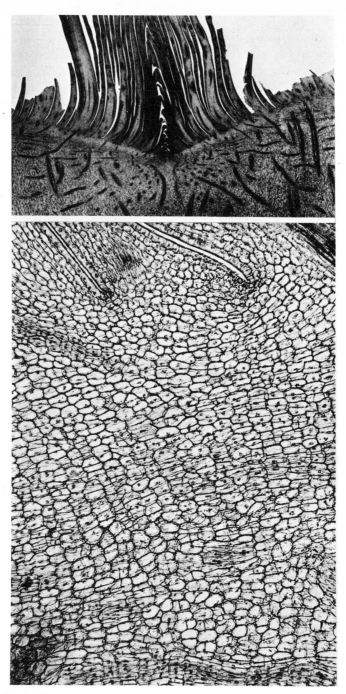

Fig. 8.1 Phormium, median L.S. through apex of vegetative shoot. A, the apex and the abruptly thickened primary axis with meristematic cap. (× 6·2.) B, detail of meristematic cap. (× 150.) Note the alternating bands of procambium and ground parenchyma from which the bulk of the stem is derived.

may continue even in the oldest basal parts of the trunk. This type of increase is due to cell-division and cell enlargement remaining prevalent generally throughout the tissues of the stem. For this reason Tomlinson (1961) applied the term *diffuse secondary thickening* to it.

Details of this process vary in different species, but in general the number of cells in the ground tissue is augmented by longitudinal divisions. They also increase in size, especially along their radial dimension. At the same time air spaces between them increase in size, though these may subsequently become filled with outgrowths from adjacent cells. The fibrous bundle sheaths also play an important part in diffuse secondary thickening. Some fibres mature early, but others remain in an undifferentiated state. Later these immature fibres increase considerably in girth and their walls undergo marked secondary thickening. Consequently, although the number of vascular bundles does not increase, the bundles themselves become considerably larger and much more widely spaced (Fig. 8.2). These processes are most active in the central tissues. The outer parts of the stem undergo increases in tangential dimensions and divide by radial walls, thus accommodating the increase in circumference (Schoute, 1912).

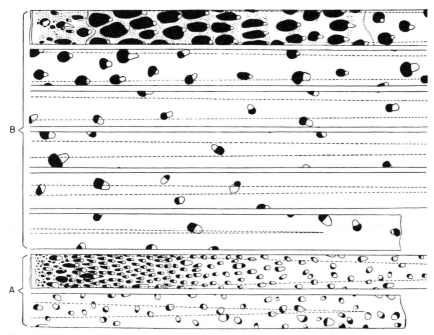

Fig. 8.2 Diagrams of sectors from younger (A) and older (B) parts of a palm trunk (*Oreodoxa*). The amount of ground tissue has greatly increased and the bundles are larger, especially the fibrovascular portion (in black). (After Schoute, 1912.)

The cambium of monocotyledons

Secondary growth due to the action of a cylindrical lateral meristem, or cambium, occurs in relatively very few monocotyledons. When a cambium is present its origin and action are distinct from any found among dicotyledons (Philipson and Ward, 1965). A rudimentary cambium is present in several monocotyledons with thick stems. Chouard (1936) figures sections through the basal part of stems with well-developed primary thickening which are clearly similar in action to the type of cambium found in monocotyledons. Cheadle (1937) found a cambium in *Veratrum*, which is a herb, but the secondary tissue derived from it is mainly parenchyma with few vascular bundles. In tree-like monocotyledons the action of the meristematic cap persists, so that a meristematic sheath envelopes the mature stem, to which it adds ground tissue and vertical bundles. This meristem is the cambium; its connection with primary thickening has been observed by several anatomists (Helm, 1936; Eckardt, 1941), and this link has been discussed earlier in this chapter. A cambium is present in the stems of the following genera (families according to Hutchinson, 1959): Liliaceae, *Aloe, Kniphofia*; Agavaceae, *Cordyline* (Fig. 8.3), *Dracaena, Dasylirion, Yucca, Agave, Furcraea, Sansevieria*; Xanthorrhoeaceae, *Xanthorrhoea*; Iridaceae, *Aristea* (and related genera); Dioscoreaceae, *Testudinaria, Tamus, Dioscorea*. The occurrence of a cambium in so many families suggests that secondary thickening may have arisen repeatedly in the monocotyledons. However, the comparison made by Cheadle (1937) shows that the cambium is of essentially the same nature in all these families. Secondary thickening of the roots of monocotyledons is much less frequent (Pfeiffer, 1926).

The activity of the monocotyledonous cambium has not been studied so meticulously as has the normal cambium of dicotyledons and gymnosperms. Schoute (1912) considered that it originates as an *etagenmeristem* (i.e. a meristem in which the individual cells are active for only a limited number of divisions, after which their function is taken over by other cells). By the time secondary parenchyma cells are being added to the cortex he considers that a true cambium has been developed with a single layer of permanent initials. The amount of secondary cortex is small, but evidently is sufficient to avoid the cumulative thickening of one tangential wall that would seem inevitable in a cambium that produced derivatives in one direction (Newman, 1956). No vascular tissue forms in these external derivatives.

In tangential view (Fig. 8.3C) the cambial initials may be more or less uniform in size and shape, or rather variable, including polygonal, fusiform, and rectangular cells (Cheadle, 1937). These variously shaped cells may be mixed together, and their diversity is retained in the mature cells of secondary conjunctive tissue to which they give rise. These differences in cell shape bear

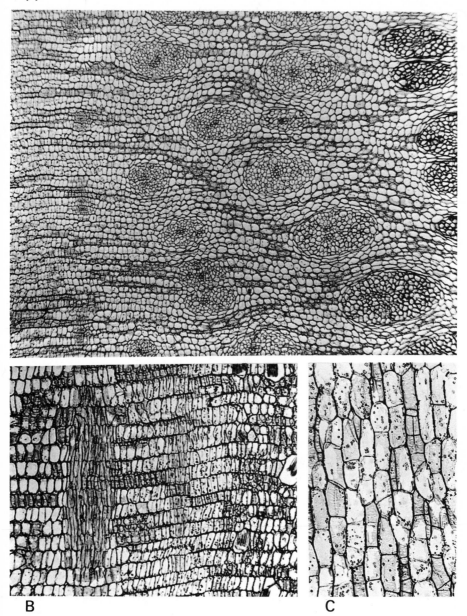

Fig. 8.3 Cordyline australis. Sections through the cambial zone. A, T.S. – the cambium is at the left of the figure with successively more mature vascular bundles towards the right. (×35.) B, R.L.S. the cortex is to the right of the figure, followed towards the left by the cambium and two procambial strands. (×70.) C, T.L.S. through the cambium. (×106.)

Fig. 8.4 Cordyline australis. A monocotyledon of tree stature.

no relation to the distinction between fusiform and ray initials in a normal dicotyledonous or gymnospermous cambium, in which the different types of initials give rise to distinct tissues. As seen in both radial longitudinal and transverse sections, the initial cells of monocotyledons are rectangular (Figs. 8.3, A and B).

The internal derivatives of the cambium divide several times before their products become differentiated into two types of cells. Most mature as secondary conjunctive tissue, the remainder being the cells from which secondary vascular bundles are derived. These latter cells divide repeatedly by longitudinal walls oriented in several directions. These groups of divisions occur in longitudinal files, resulting in continuous procambial strands running vertically in the maturing conjunctive tissue (Fig. 8.3B). From these strands are differentiated the phloem and xylem elements of the new vascular bundles. The phloem and parenchyma elements do not elongate significantly during their differentiation, remaining approximately the same length as the cambial initials. The tracheary elements, on the other hand, elongate to as much as forty times their original length by means of intrusive growth. Very few cells in each tier of the procambial strands differentiate into tracheary elements, but as a result of their elongation several come to be present in a transverse section of a mature strand.

There can be no doubt that the presence of a cambium in dicotyledons and gymnosperms is an important factor in the development of the tree habit. Among other factors, it develops the mechanical tissues needed to support an enlarging crown of branches. The dendroid habit occurs in three groups of the monocotyledons, namely, the screw-pines, the palms, and the Agavaceae (Fig. 8.4). In the screw-pines (*Pandanus*) the stems lack secondary thickening, but the weight of the crown is partly supported by the stilt and prop roots. In palms a trunk is produced without the intervention of a cambium, so that only in the Agavaceae is a cambium present, and it is quite unlike any found among the dicotyledons. Increase in the diameter of stems during their primary growth appears to be the general situation in monocotyledons. From this condition, diffuse secondary thickening may be derived by an extension of the divisions normally present in internodes during extension and differentiation. The relationship between primary thickening and the cambium has already been described. The broadly based monocotyledonous leaf is supplied by many bundles which form at intervals after leaf initiation. The main bundles, which form before the action of the thickening meristem has displaced the leaf primordia from near the stem centre, have their courses greatly distorted by its action. Later-formed subsidiary bundles will pursue a more straightforward course. This method of primary thickening, therefore, not only results in an arrangement of bundles inappropriate to a normal cambium but also provides the basis of a distinctive method of secondary increase.

REFERENCES

ARBER, A. (1917). On the occurrence of intrafascicular cambium in monocotyledons. *Ann. Bot.* **31**, 41–5.

— (1918). Further notes on intrafascicular cambium in monocotyledons. *Ann. Bot.* **32**, 87–9.

— (1919). Studies in intrafascicular cambium in monocotyledons. III and IV. *Ann. Bot.* **33**, 459–65.

— (1925). *Monocotyledons, a Morphological Study*. Cambridge University Press, Cambridge.

CHEADLE, V. I. (1937). Secondary growth by means of a thickening ring in certain monocotyledons. *Botan. Gaz.* **98**, 535–55.

CHOUARD, P. (1936). La nature et la rôle des formations dites 'secondaires' dans l'édifications de la tige de Monocotylédones. *Bull. Soc. Bot. Fr.* **43**, 819–36.

CLOWES, F. A. L. (1961). *Apical Meristems*, Blackwell, Oxford.

ECKARDT, T. (1941). Kritische Untersuchungen über das primäre Dickenwachstum bei Monokotylen, mis Ausblick auf dessen Verhältnis zur Sekundären Verdickung. *Bot. Arch.* **42**, 289–334.

HELM, J. (1936). Das Erstarkungwachstum der Palmen und einiger anderer Monokotylen, zugleich ein Beitrag zur Frage des Erstarkungwachstums der Monokotylen überhaupt. *Planta* **26**, 319–64.

HOLTTUM, R. E. (1955). Growth-habits of monocotyledons – variations on a theme. *Phytomorphol.* **5**, 399–413.

HUTCHINSON, J. (1959). *The Families of Flowering Plants*, Vol. II, Monocotyledons, Oxford University Press, Oxford.

NEWMAN, I. V. (1956). Pattern in meristems of vascular plants. 1. Cell partition in living apices and in the cambial zone in relation to the concepts of initial cells and apical cells. *Phytomorphol.* **6**, 1–19.

PFEIFFER, H. (1926). Das abnorme Dickenwachstum, in LINSBAUER, K., *Handbuch der Pflanzenanatomie, Bd IX*, Bornträger, Berlin.

PHILIPSON, W. R. and WARD, J. M. (1965). The ontogeny of the vascular cambium in the stem of seed plants. *Biol. Rev.* **40**, 534–79.

SCHOUTE, J. C. (1912). Über das Dickenwachstum der Palmen. *Ann. Jard. Bot. Buitenzorg.* **26**, 1–209.

TOMLINSON, P. B. (1961). *Anatomy of the Monocotyledons, II. The Palmae*, Oxford University Press, Oxford.

ZIMMERMANN, M. H. and TOMLINSON, P. B. (1967). Anatomy of the Palm *Rhapis excelsa*, IV. Vascular development in apex of vegetative aerial axis and rhizome. *J. Arnold Arb.* **48**, 122–42.

— and — (1968). Vascular construction and development in the aerial stem of *Prionum* (Juncaceae). *Am. J. Botany* **55**, 1100–9.

9
Cambial activity

The structure and activity of the vascular cambium is not uniform, but shows great variation according to the differing genetic constitutions of plants and differences in internal conditions and the external environment. This chapter is concerned with seasonal changes within the cambium, and particularly with temporal variations in the frequency of periclinal division and the factors which bring about such variations.

Periclinal division in the cambium leads to the formation of secondary vascular tissues, and the history of cambial activity can be followed in these tissues. The radial growth of trees and the factors influencing it, especially with reference to wood production, have been the subject of extensive and intensive study for centuries. Useful reviews of publications in this field include those of Priestley (1930), Wareing (1958), and Reinders-Gouwentak (1965) on the physiology of cambial activity, Grossenbacher (1915) and the literature review in Ladefoged (1952) on the periodicity of wood formation, Glock (1941, 1955) on growth rings and climate and Studhalter (1955) and Studhalter *et al.* (1963) on tree growth.

The periodicity of cambial activity

Most woody plants grow periodically rather than continuously, and this is true of both apical and radial growth. In temperate regions such periodicity is clearly correlated with the change in seasons, with growth taking place in spring and summer and being followed by a period of dormancy which extends to the following spring. In tropical regions growth may be continuous or intermittent. Periodicity in the tropics may be less clearly correlated with seasonal changes in environmental conditions than in temperate regions.

The normal seasonal pattern of cambial activity in the Temperate Zones is reflected in the formation of annual rings in the wood; the boundaries between annual growth increments or rings are clearly demarcated due to the formation of morphologically distinct late wood towards the end of one growing season and early wood at the beginning of the next. Multiple growth, in which there is more than one growth period in a year, is not uncommon in temperate regions, and may lead to the formation of false rings (Fig. 9.7). It is usually brought about by the reduction or cessation of cambial activity during the normal growing season in response to environmental conditions, such as temporary drought or low temperatures, which inhibit growth.

In the tropics, where environmental conditions are usually less limiting, the growing season tends to be longer than in temperate regions. But, even when conditions are ideal, growth is not always continuous. Koriba (1958) found that in Malaysia, where growth conditions are ideal and nearly uniform throughout the year, only about 15% of species show continuous radial growth, although the majority are evergreen. Only 43% of Amazon rain-forest trees show no growth rings (Alvim, 1964), although the figure estimated by Chowdhury (1961) for rain-forest trees in India is higher (75%). Even when growth is continuous, the rate of growth may vary periodically (Alvim, 1964). Because tropical climates are often regarded as non-seasonal, intermittent growth in tropical species has frequently been ascribed to internal growth rhythms. Alvim found that growth periodicity is greater where environmental factors, such as rainfall, temperature, and day length, vary seasonally and he considers that more intermittent growth in the tropics is caused by environmental factors than is commonly realized.

Growth rings in tropical species may not be annual, although even under rain-forest conditions annual rings may be formed. Multiple growth is common, and the number of rings formed in a given period of time may be constant or variable, depending on whether environmental conditions vary consistently or erratically. Bailey (1944) found that annual rings may be obligate, with extra rings being facultative according to environmental conditions. In arid regions only one growth ring may be formed over several years, with a burst of cambial activity apparently occurring as water becomes available (Holtermann, 1902).

Seasonal changes in the cambium

The dormant cambium consists of a narrow zone of radially flattened cells. The cell walls are relatively thick, and in tangential view have a beaded appearance because of alternate thickened areas and deeply depressed primary pit-fields. The cytoplasm is relatively dense. The cambial zone is relatively narrow, although the number of cells in a radial file varies. According to Bannan (1962), in conifers there are one to five and usually two or three cells consisting of the initial and a varying number of xylem mother cells

(Fig. 9.1A). Larger numbers of cells have been reported, and this seems to be related at least in part to whether undifferentiated or partly differentiated phloem cells overwinter next to the cambial initials and whether they are regarded as part of the cambial zone. The dormant cambium is usually narrower in dicotyledons than in conifers (Ladefoged, 1952).

Fig. 9.1 Transverse sections of the cambium of *Thuja occidentalis*. A, the dormant cambium in February; B, swelling of the cambial cells on reactivation in mid-April. (From Bannan, 1962.)

The first sign of reactivation of the cambium in spring is usually the phenomenon known as swelling. The dark, shrunken cells of the cambium take on a lighter, turgid appearance. They expand radially, the radial walls become thinner, the protoplasts become less densely granular, and the nuclei enlarge (Fig. 9.1B). Priestley (1930) suggests that swelling represents the resumption of a sol state by water uptake after the frost-resistant gel state of the winter cambium. Cambial swelling does not occur in all species; the cells do not enlarge noticeably in *Pyrus communis* and *P. malus* L. (Evert, 1960, 1963), *Robinia pseudoacacia* (Derr and Evert, 1967) and *Populus tremuloides* Michx. (Davis and Evert, 1968). Alfieri and Evert (1968) found that in *Pinus* no general expansion of the cambium precedes the onset of division, but rather each cambial cell expands independently before dividing. Swelling begins at the base of the buds (Ladefoged, 1952) and may spread very rapidly over the whole tree (Wareing, 1958) or take up to several weeks to reach the base of the trunk in some diffuse-porous species (Wilcox, 1962).

The swelling of the cambium is followed by the onset of periclinal division and the production of new secondary xylem and phloem cells (Fig. 9.2). The location of the first divisions in the radial files of cells in the cambial zone varies; they may be regularly distributed across the zone (Evert, 1963; Derr

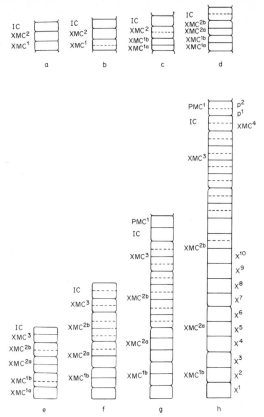

Fig. 9.2 Series of diagrams showing sequence of cell divisions following reactivation of the cambium in spring. Divisions are shown as beginning in xylem mother cells, extensive re-division of which results in rapid production of xylem-destined cells. Less frequent divisions in the tier of initial cells yield new xylem mother cells on the inside and phloem mother cells towards the outside. Symbol IC denotes initial cell; XMC, xylem mother cell; PMC, phloem mother cell; X, differentiating xylem element; and P, phloem element. (From Bannan, 1962.)

and Evert, 1967), or they may be more common in the cells next to the xylem (Bannan, 1955). Bannan found that behaviour varied in different radial files, with the first division occasionally occurring in the initial or a xylem mother cell not adjacent to the xylem and with the mother cell next to the xylem occasionally swelling and differentiating without undergoing division.

The position in the tree of the first periclinal divisions is usually at the base of the buds, and from here the onset of division spreads basipetally down the branches and trunk. The rate of spreading varies among diffuse-porous and ring-porous dicotyledons and conifers. In diffuse-porous dicotyledons the spread is relatively slow, and division at the base of the trunk may not begin until several weeks after it begins in the twigs (Wareing, 1958; Wilcox, 1962). In ring-porous dicotyledons it is more rapid and usually too rapid to detect a time lapse (Wilcox, 1962), although Tepper and Hollis (1967) found that it was slower, moving at a maximum of 6 cm a day, in 2-yr-old *Fraxinus* seedlings. Conifers tend to show an intermediate condition, with spread usually taking about a week (Wilcox, 1962).

With the onset of cambial activity the bark may be readily separated from the wood, since the radial walls of the cambial cells, and later its differentiating derivatives, are weak and easily broken. This phenomenon is known variously as sap-peeling or slipping or peeling of the bark. Its use to detect cambial activity is discussed by Wilcox (1962). Priestley (1930) considered that slipping occurred along the plane of the cambium and was associated with cambial swelling. While slip may occur through the cambium before differentiation begins, at later stages it generally takes place through the differentiating xylem in the region where the cells have expanded but are still very thin-walled (Bailey, 1943; Evert, 1960; 1963).

Once initiated, periclinal divisions usually continue for some months. The length of the season for radial growth is generally 60–100 or more days in North Temperate hardwoods, but it may be as little as four to six weeks or as much as six months. It tends to shorten in higher latitudes and lengthen towards the equator, and is generally longer in conifers than in dicotyledons (Studhalter *et al.*, 1963). Cambial activity in the North mid-Temperate Zone usually begins in April or May and ceases in August or September, but may vary with species, vigour of growth, and environmental conditions. Annual radial growth in temperate regions usually shows the typical sigmoid growth curve. There is generally a rapid increase in activity at the beginning of the season, followed by a levelling off and a gradual decline. The grand period of growth, in which cambial activity is at a high level, is usually sustained for a considerable time, although Catesson (1964) found it to be of very short duration in young branches of *Acer pseudoplatanus*. Daubenmire and Deters (1947) found that the cambium tends to be most active in the early part of the grand period of growth in evergreen conifers and in the later part in deciduous dicotyledons. The rapid increase in activity in spring may be preceded by a period of low-level activity (Evert, 1960, 1963); this appears to be associated with the production of phloem before the build-up in xylem production. The decline in cambial activity is usually considered to be a gradual process, but in dicotyledons at least it may be quite rapid (Evert, 1960, 1963; Catesson, 1964; Derr and Evert, 1967), although an initially rapid decline may slow to a gradual one before division finally ceases (Catesson, 1964). The usual sigmoid curve for radial growth is dependent on, and may be modified by, conditions in the external environment.

The activity of the cambial-ray cells tends to lag behind that of the fusiform cells. Evert (1963) found that in *Pyrus malus* periclinal divisions begin later, reach their greatest frequency later, and stop sooner in the ray cells. Early in the season, when only the fusiform cells are dividing, the ray cells keep up with the rate of expansion by elongating radially. Catesson (1964) found that anticlinal divisions in 2-yr-old branches of *Acer pseudoplatanus* similarly extend over a shorter season and reach their highest frequency later in ray initials than in fusiform initials.

The pattern of radial growth is paralleled by the width of the cambial zone, which increases, remains constant, and declines with the rate of cell production. Wilson (1966) found that the variation in the number of cells across the cambial zone seems to reflect a balance between the rate of cell division and the rate of differentiation of new derivatives. Division proceeds faster than differentiation early in the season, resulting in a widening of the cambial zone. Once differentiation begins, a balance is soon established, and the width of the zone stays more or less constant until the rate of division slows, when differentiation proceeds faster than division and the cambial zone narrows. The balance between division and differentiation is achieved in both vigorous trees with relatively wide cambial zones and slow-growing trees with relatively narrow ones, indicating that the rate of differentiation must vary with the overall rate of production.

During the surge of radial growth the most actively dividing cells are those towards the xylem. Bannan (1964) found that in *Pseudotsuga*, in an average ring 3 mm wide, the succession of functional fusiform initials in a radial file produces no more than five or six xylem mother cells, and these by redivision produce the 100 or so tracheids of the annual ring. The phloem mother cells usually redivide once or occasionally differentiate without dividing.

Most of the anticlinal divisions in the cambial initials, leading to the increase in girth of the cambium, occur towards the end of the growing season in mature stems (Bannan, 1950, 1951, 1955, 1960, 1964; Evert, 1961, 1963; Derr and Evert, 1967; Cumbie, 1967). The greatest elongation of fusiform initials occurs mainly towards the end of the season of radial growth (Bannan, 1951) or between the cessation of radial growth and the production of xylem mother cells the following spring (Evert, 1961). The cambium is thus prepared for the increase in circumference which will be required during intensive xylem production in the spring. This must be achieved by the tangential expansion of the cells which have recently undergone anticlinal division and elongation. Evert (1961) found that the frequency of anticlinal division in pear trees is two or three times greater every other year. This is probably related to a biennial fluctuation in radial growth, caused by fruiting (see p. 133). The exact timing of anticlinal division in relation to periclinal division has been studied by Bannan (1955) in *Thuja*. Some anticlinal divisions occur while xylem and phloem are still being produced, some just before the last or next to last periclinal division of the season, and a few after radial growth has ceased. In very vigorous or very young stems, where the rate of periclinal division is relatively high compared with that of normal mature stems, anticlinal division may occur throughout the season of radial growth (Bannan, 1950, 1960; Catesson, 1964; Cumbie, 1967; Derr and Evert, 1967). The rate of loss of fusiform initials from the cambium tends to increase through successive quarters of the annual ring, but this may be associated with the fact that the xylem is laid down more slowly towards the end of the season.

The annual increment of phloem varies less from year to year and between vigorous and slow-growing trees than that of xylem (Artschwager, 1945; Bannan, 1955; Wilson, 1964). Vigorous conifers may produce ten times as much xylem as phloem, while in slow-growing ones the amounts approach equality, with the amount of phloem remaining relatively constant (Wilson, 1964). This difference is undoubtedly related to the fact that phloem mother cells usually redivide only once before differentiation, while the xylem mother cells may undergo several successive divisions, with the number rising in vigorous trees and at the height of xylem production.

Although the production of xylem is not usually entirely separate in time from phloem production, there is generally a predominance of one or the other. In *Thuja* (Bannan, 1955) the seasonal growth curves for xylem and phloem are quite different, the xylem curve reaching a high, early peak and later tapering off, while the phloem one begins later and continues at a more or less uniform level throughout the season, with sometimes a slight rise towards the end (Fig. 9.3). An early predominance of xylem production has been reported by many other workers (Raatz, 1892; Brown, 1915; Gill, 1932; Artschwager, 1945, 1950; Esau, 1948; Grillos and Smith, 1959). In some cases, however, most of the phloem is formed early in the season, either before the xylem (Evert, 1963) or during the surge of xylem production (Knudson, 1913) or both (Derr and Evert, 1967; Tucker and Evert, 1969). In *Acer pseudoplatanus*, Catesson (1964) reports that an early stage in which mainly phloem is produced is followed by a period of predominantly xylem production and then a second period of phloem production. An increase in phloem production late in the season has also been reported by Gill (1932).

Phloem formation has been reported to begin before, after, or simultaneously with xylem formation and to cease later or at the same time. But, from the multitude of reports, little can be determined with accuracy about behaviour within the cambium. The time of formation of xylem and phloem is not necessarily an accurate reflection of the time of xylemward and phloemward cell production in the cambium. The formation of the vascular tissues

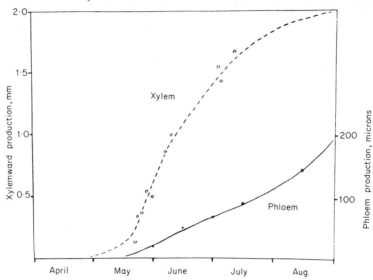

Fig. 9.3 Graph showing the march of xylem and phloem production during the growing season for a hypothetical white-cedar tree having an annual xylem increment of 2 mm. (From Bannan, 1962.)

involves both division and differentiation, and these processes have not usually been distinguished in the literature. In many species immature phloem elements overwinter and complete maturation at the beginning of the following season. These may or may not have been regarded as part of the cambial zone. This problem is discussed by Evert (1963). Studies of the cambium indicate that the first divisions may be in the cells near the xylem (Bannan, 1955; Grillos and Smith, 1959), adjacent to the phloem (Tucker and Evert, 1969), or anywhere across the cambial zone (Evert, 1963; Derr and Evert, 1967), and in the early part of the season the majority of divisions in a radial file may produce cells towards the xylem (Bannan, 1955) or the phloem (Evert, 1960, 1963; Catesson, 1964; Derr and Evert, 1967; Tucker and Evert, 1969). Whether xylem or phloem is produced may depend on the relative concentrations of different hormones (Digby and Wareing, 1966a, and see Chapter 10).

The work of Evert and his associates (Evert, 1960, 1962, 1963; Davis and Evert, 1966; Derr and Evert, 1967; Davis, 1968; Davis and Evert, 1968; Tucker, 1968; Tucker and Evert, 1969) has revealed a correlation between the type of wood porosity and the pattern of radial growth. In ring-porous species studied, phloem and xylem differentiation were found to begin virtually simultaneously. In diffuse-porous species phloem differentiation begins a month to six weeks before xylem differentiation. An exception is *Tilia americana* L. (Evert, 1962), in which both begin at the same time; however, this species has overwintering phloem elements which are reactivated about a month earlier. *Pinus* is similar to diffuse-porous dicotyledons, with differentiation of phloem preceding that of xylem (Alfieri and Evert, 1968). In none of the species studied did xylem begin to differentiate before phloem.

Growth correlations

Since absorption, conduction, and photosynthesis are dependent on one another, it is clear that there must be a close relationship among the different growing regions of the plant. While the relationship between terminal- and radial-shoot growth has been studied extensively, more comprehensive studies which include root growth seem to be very rare. Kienholz (1934) compared data on leader growth, needle growth, cambial activity, and root elongation in red and white pine (Fig. 9.4). Root elongation begins first, as might be expected, since water is essential for growth, and continual root extension is

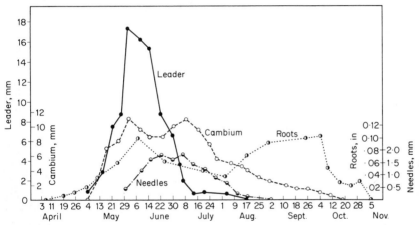

Fig. 9.4 Interrelation of the different growth activities of pine. (From Kienholz, 1934.)

necessary for water uptake (Kramer and Coile, 1940). The surge of growth in roots, leaders, and cambium occurs at about the same time, probably because water and stored food are abundant. After this the times of maximum activity are different for the different growing regions, suggesting that a balance is achieved among them in an internal competition for growth materials. Reed and MacDougal (1937), studying root and shoot elongation and cambial activity in orange trees, obtained similar results.

It has been known for centuries that a correlation exists between terminal and radial growth in the shoot, but the data have been contradictory, and until recently apparently inexplicable. However, recent work on the role of hormones in growth (see Chapter 10) has led to greater, although by no means complete, understanding of the relationship between the activity of the apical meristem and that of the vascular cambium.

That a relationship exists between bud break and the onset of cambial division has long been recognized. Priestley (1930) was the first to elucidate this. After a thorough examination of the literature he concluded that in dicotyledons cambial activity normally begins at the bases of the apical buds and that, while division might occasionally occur in the trunk before the twigs, there was no evidence of activity beginning first in areas not closely associated with growing buds. He found the situation in conifers to be somewhat different, with cambial division in many species habitually beginning separately in the main axis and branches, although always at the bases of breaking buds. However, in view of Hartig's observations that in some species budless stumps increase in girth for years, while in other species they do not, the difference apparently being attributable to differences in supplies of stored carbohydrates, Priestley concluded that the connection between buds and the initiation of cell division in the cambium is not always obligate in conifers. Wareing (1958) reports that the cambium of conifers may become active in the absence of buds but produces abnormal wood. Whitmore and Zahner (1966) found that the cambium of red pine becomes active even if the buds are removed before terminal growth begins. It is generally accepted that the stimulus which initiates cambial division is auxin produced in the reactivated buds. Wort (1962) states that the dormant cambium of many species, cultivated *in vitro*, resumes activity in the absence of any added growth-promoting substance when exposed to temperatures of about $25°C$, and suggests that reserves of auxin may be present in tissues adjacent to the cambium and be liberated by temperature increases. This could account for some reports of initiation of cambial division in the absence of buds.

The time of the initiation of cell division in the cambium in relation to the time of bud break is different in diffuse-porous and ring-porous species, tending to be considerably earlier in the latter. Ladefoged (1952) found that cambial activity begins in diffuse-porous species up to a week before bud break where the shoot extends considerably before the buds burst and otherwise simultaneously with or at the most two days before bud break. In ring-porous species it begins one to nine days before and in conifers five to fifteen days before bud break. Different timing has been reported in tropical trees (Alvim, 1964; Reinders-Gouwentak, 1965), with cambial activity commencing later in relation to bud break. In *Michelia* (diffuse-porous) and *Melia* (semi-ring-porous) in India, cambial activity does not begin until well after the cessation of the first of two periods of terminal growth (Chowdhury and Tandan, 1950). In the semi-arid conditions of the Israeli maquis, cambial activity in dicotyledonous trees does not begin until four to six weeks after bud extension, probably because of a shortage of stored food at the commencement of terminal growth (Fahn, 1953, 1955).

The onset of cambial division progresses basipetally except in the current year's shoot, where it progresses acropetally, since radial growth does not occur until the primary tissues have ceased to elongate. It spreads rather slowly in diffuse-porous species, so that the tree is usually in full leaf before the entire cambium is dividing, and very rapidly in ring-porous species, with the entire cambium often undergoing division while the buds are still in an early stage of expansion. The rapid spread of activity in ring-porous species may be due to the presence in the cambium or cortical tissues before bud break of a hormone precursor which is converted to auxin at all levels in the tree at the time of bud reactivation (Wareing, 1951b; Digby and Wareing, 1966b). Reports of non-basipetal spreading of cambial activity may be due to such a precursor, liberated by high temperatures (Wort, 1962). Digby and Wareing (1966b) have shown that the pattern of spreading of cambial activity is correlated with auxin gradients in both diffuse-porous and ring-porous species.

The relationship between the rates of terminal and radial growth varies. Maximum radial growth often occurs at the same time as maximum terminal growth (Avery et al., 1937; Priestley, 1930), but Alvim (1964) reports that in Cacao in Costa Rica cambial activity is reduced or ceases at the time of or just after intensive tip flushing, and is greatest when limited tip flushing is combined with favourable conditions for photosynthesis. The amount of radial and terminal growth in a season usually shows a positive correlation, although instances in which there is no correlation or even negative correlation have been reported (Studhalter et al., 1963). It is possible that the radial growth rate increases with the terminal growth rate due to a rise in auxin levels, provided that other growth materials, such as water and carbohydrates, are not limiting.

In species in which multiple growth occurs, the number of rings formed may be the same as the number of periods of terminal growth, or it may be smaller or, more rarely, greater (Studhalter et al., 1963). False rings are associated with temporary cessation of terminal growth and the resultant drop in auxin synthesis. However, a false ring is formed only if the interval between periods of terminal growth is long enough for the auxin level to become low enough for late wood to be formed, and if the following period of terminal growth is vigorous enough to raise auxin levels sufficiently for the production of early wood. Moreover, the changes in the level of auxin production may not be of sufficient intensity or duration to be felt in the lower regions of the tree, so that false rings may be restricted to the upper part of the crown or even to the juvenile wood. This may account for the many reported instances of false rings failing to accompany multiple tip growth

(Larson, 1962). Wareing (in Zimmerman, 1964, p. 493) points out that in European oak two tip flushes do not necessarily lead to the formation of two growth rings at the base of the first shoot, and suggests that while the bud is quiescent the cambium is kept active by auxin from the mature leaves. Chowdhury (1958) reports that in India trees show from two to four periods of extension growth but only one period of radial growth. In some species with two periods of extension growth the cambium does not become active at all during the first of these (Chowdhury and Tandan, 1950). The occurrence of more rings than periods of terminal growth is apparently not common (for reported instances, see Studhalter *et al.*, 1963), except in ponderosa pine, in which two rings often accompany one period of tip growth. The explanation may lie in an internal competition for growth materials, such as water and carbohydrates, or in a fluctuating rate of tip growth, which alters auxin levels in the cambium and through this affects the rate of cambial activity and the type of wood formed.

The cessation of cambial activity and of terminal growth are fairly closely correlated in many diffuse-porous species (Avery *et al.*, 1937; Wareing, 1958). In some ring-porous species and conifers cambial activity continues after the cessation of terminal growth (Wight, 1933; Wareing, 1958; Wareing and Roberts, 1956). The mature leaves of some species produce small amounts of auxin, and it has been suggested by Wareing (1956, 1958) and Wareing and Black (1958) that, since the continued activity of the cambium is not prevented by bud removal (Wareing, 1951a; Wareing and Roberts, 1956), the stimulus may come from auxin produced in the mature leaves. Digby and Wareing (1966b) have shown that this theory is correct. Auxin is produced after terminal growth ceases in the mature leaves of ring-porous species, but little is produced under these conditions in the leaves of diffuse-porous species.

The activity of the cambium is affected by reproductive growth as well as by vegetative tip growth. The developing reproductive structures produce auxin, and if other factors are not limiting this stimulates cambial activity. In woody species which flower before leaf production, flowering initiates cambial activity, although the activated area may not extend very far downwards from the flowers if insufficient amounts of auxin are produced. Gill (1933) studied the influence of flowering in four male catkin-bearing trees. In species in which the catkins are already fully formed and merely open in spring there is no cambial activity, but where the catkins differentiate in spring, wood and phloem are formed in the axis of the catkin and some phloem in the twig below. Reinders-Gouwentak and van der Veen (1953) confirmed these observations and found that cambial activity is promoted in a more conspicuous manner in fertilized female catkins. Reinders-Gouwentak (1965, pp. 1096–8) discusses reports of the effects of flowering in annuals and biennials and the spurs of some fruit trees. Flowering is accompanied by a reduction of cambial activity, but this is because flower initiation has stimulated the cambium of flowering shoots so that it produces more rapid secondary thickening than the cambium of vegetative shoots. Once the ultimate diameter of the flowering stem is attained, cambial activity ceases, and the cambium may be inactive during flowering. The reproductive stem ages more rapidly than the vegetative one.

A quite different relationship between reproductive and cambial activity has been reported for many woody plants. It appears to depend primarily on carbohydrate rather than auxin levels. Fruiting in orchard trees and heavy seeding in forest trees may reduce the level of cambial activity markedly (Antevs, 1925; Glock, 1955; Huber and Jazewitsch, 1956; Mott *et al.*, 1957), and this effect may extend to the following year. It may affect tip growth as well (Morris, 1951). Tingley (1936) found that two growth rings may be formed in heavy fruiting years and only one in off years in apple trees. Heinecke (1937) estimated that 35% of total carbohydrate (about the same amount as is used in the production of structural tissues) is required for fruit production in apple trees. A biennial fluctuation in radial growth due to decreases associated with fruiting may occur in apple and pear trees. Decreases in cambial activity associated with reproduction appear to be caused by competition for carbohydrates. Apart from the direct effect of carbohydrate shortage on cambial growth, a concomitant decrease in terminal growth may lower auxin levels and thus affect the cambium indirectly. In addition, there is considerable evidence that translocatable carbohydrates are required for the downward movement of growth hormones (Kozlowski, 1962).

Factors influencing cambial activity

Cambial activity is affected by a variety of factors both inside and outside the plant. These include hereditary constitution, physiological processes, and environmental factors. They form a complex of interrelated influences, and thus the effects of any one factor are often very difficult to assess. Growth pattern, like other characters, is the product of the interaction of genes and environment, and will vary only within genetically fixed limits of tolerance. Thus some tropical trees grow periodically, although conditions for growth are apparently ideal throughout the year. Geographical races of a species when grown together may show different responses to such external factors as photoperiod, and different species and different individuals of the same species growing on the same site under the same environmental conditions may have different patterns of growth.

Within the limits set by the genotype, cambial growth is dependent on the availability of water, minerals, carbohydrates, and hormones. Any factor, internal or external, which limits the availability to the cambium of any of these growth materials may become limiting for radial growth. The environmental factors which have been most studied in relation to cambial activity are rainfall and temperature, but others include soil moisture, light intensity, photoperiod, competition, crown size, oxygen and carbon dioxide levels, wind, defoliation, topography, latitude, and altitude.

It should be emphasized that a single environmental factor may affect many different processes in the plant, and in addition may affect, be affected by, and interact with other external factors. This point is well illustrated by the effects of water on cambial activity and the relationship of rainfall to radial growth. The role of water in tree growth is discussed by Kramer (1962). Water affects cambial activity directly, since loss of turgor decreases cell enlargement. There are many indirect effects from within the plant, since water is required for the growth of other meristems, which affects cambial activity, and for metabolic processes such as photosynthesis and nitrogen metabolism and processes such as translocation and salt absorption. The water balance within the plant is controlled by the relative rates of water absorption and water loss. Water loss is affected by such plant factors as leaf area, leaf structure, and stomatal behaviour, and by environmental factors such as solar radiation, temperature, humidity, and wind. Water absorption depends on the rate of water loss, the extent and efficiency of root systems, the

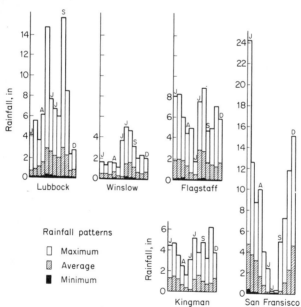

Fig. 9.5 Maximum, average, and monthly rainfall of Lubbock, Texas; Winslow, Flagstaff, and Kingman, Arizona; and San Francisco, California. (From Glock and Agerter, 1962.)

presence of soil moisture, soil aeration, soil temperature, and factors affecting the availability of soil moisture, such as soil-moisture tension and concentration of the soil solution. Soil moisture during the growing season depends on a number of factors, including amount of rainfall, proportion of rainfall occurring before and during the growing season and its distribution during the growing season, the frequency and heaviness of falls, topography and its relation to run off, and the character of the soil and its ability to store water. Ability of the soil to store water in turn depends on climate, the depth and nature of the water table, local geology, and the root habits of trees, while rainfall is affected by such factors as temperature and topography and itself affects light, temperature, and humidity.

In spite of the complexity of the relationship between them, rainfall and cambial growth are often closely correlated. Growth curves more often show a relationship to rainfall than to any other single factor (Glock, 1955), not because the relationship is a simple one but because water is the most common limiting factor for growth, particularly in warmer climates. Glock and Agerter (1962) found that under three different rainfall regimes (Fig. 9.5) three characteristic patterns of radial growth occur. The California pattern shows regular growth increments and occurs under a winter rainfall regime in which there is a rainy winter season and a dry summer season. Soil water is at its maximum at the beginning of the growing season and diminishes steadily, sometimes becoming low enough to bring about the cessation of radial growth. The West Texas pattern is characterized by partial growth layers (lenses), multiple growth, and high degree of radial variability in thickness. It occurs under a summer rainfall regime in which the bulk of the rainfall comes after the first burst of growth and brief wet spells during the summer alternate with dry spells which may be short or prolonged and intense. The two contrasting patterns are evident in *Agathis* from a California type of rainfall regime in New Zealand and *Araucaria* from a West Texas type of rainfall regime in eastern Australia (Fig. 9.6). Details of a typical example of the West Texas pattern are shown in Fig. 9.7. The North Arizona pattern typically shows alternating groups of growth layers of the former two patterns (Fig. 9.8) and occurs under a regime of summer and winter rainfall.

Walter (1963) discusses an instance in which rainfall is a very inappropriate measure for soil moisture. Degree of run-off tends to increase as the density of vegetation decreases, and even in nearly flat deserts run-off is high if the vegetation cover is low. Water accumulates in depressions, runnels, and wadis and penetrates quite deeply into the soil; the upper layer of soil dries out, but moisture may be conserved for years in deeper layers. The possible effect of such run-off is apparent in *Acacia* species which grow in wadis in the Negev desert and show continuous cambial activity although the annual rainfall is less than 5 mm (Fahn, 1959).

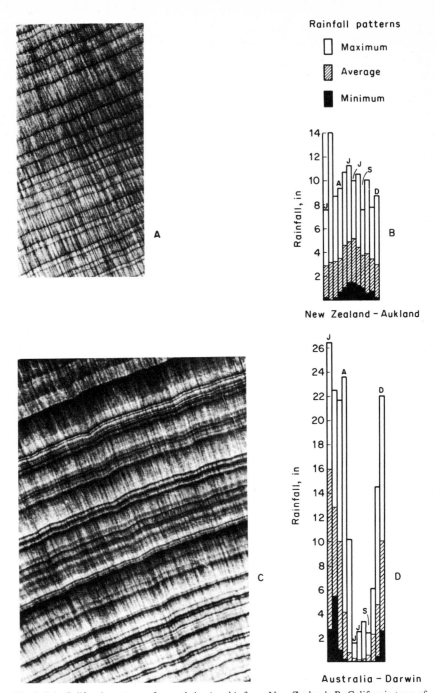

Fig. 9.6 A, California pattern of growth in *Agathis* from New Zealand: B, California type of rainfall regime: monthly rainfall in Auckland, New Zealand: C, West Texas pattern of growth in *Araucaria* from eastern Australia: D, West Texas type of rainfall regime: monthly rainfall in Darwin, Australia. (From Glock and Agerter, 1962.)

Fig. 9.7 The West Texas pattern of growth: enlarged transverse section of Arizona cypress (*Cupressus arizonica* Greene) branch from a tree growing near Lubbock, Texas. The section includes from centre to bark: 1937, central growth layer; 1938, natural frost effects of 7–9 April, one thick growth layer, two thin growth layers, one long lens; 1939, spots of frost effects, one thick growth layer, one lens-like growth layer, one thin growth layer; 1940, frost effects, one thick growth layer, two thin growth layers, lensing; 1941, a complete growth layer, branch cut from tree 11 October. (From Glock and Agerter, 1962.)

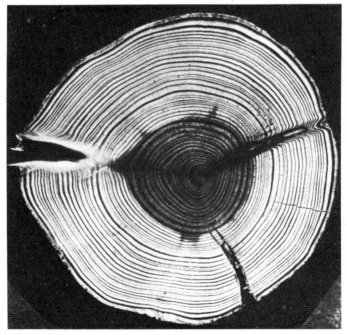

Fig. 9.8 The Northern Arizona pattern of growth in Douglas fir, with a rather rapid alternation of groups of uniform and variable growth layers. (From Glock and Agerter, 1962.)

Soil moisture is often limiting for radial growth towards the end of the season (Freisner, 1941; Pearson, 1950; Fritts, 1958; Kozlowski *et al.*, 1962), and this effect may be intensified by high summer temperatures which increase the transpiration rate. Studies of radial growth in trees growing on sites with different amounts of soil moisture (Fritts, 1958; Fraser, 1962) have shown that radial growth begins sooner and finishes later when the soil is moist and well drained than when it is either dry or very wet. On very wet soil cambial activity may be limited because the soil is poorly aerated and the roots are unable to function properly. Fraser (1956) found that cambial activity begins later in Ontario, Canada, on wetter sites because the thick organic horizon delays the warming of the soil in spring. Periods of drought may bring about the temporary cessation of cambial activity and the formation of false rings (Ladefoged, 1952; Glock, 1955; Glock and Agerter, 1962; Kennedy and Farrar, 1965). The effect of drought depends, however, on soil conditions and the structure of the root system. Ladefoged found that during a dry period cambial activity decreased and a false ring was formed in a shallow-rooted species but not in deep-rooted species nor in trees growing on swampy sites.

Temperature is most important as a limiting factor for cambial activity in high latitudes, at high altitudes, and at the beginning of the growing season in temperate regions. It affects cambial activity directly and also indirectly through its influence on other activities in the plant and on other environmental factors. Mikola (1962) studied the effect of temperature on radial growth in Finland, where temperature decreases gradually northwards while other conditions remain fairly uniform. He found that it is there the dominant factor in radial growth, and its decisive influence increases northwards, being greatest at the northern timber-line. Even in the tropics there may be a positive correlation between cambial activity and air temperature, and at a latitude of only 14° 30′ S temperature is one of the main factors controlling growth periodicity (Alvim, 1964). In temperate regions temperature is the environmental factor which usually determines the time of reactivation of the cambium and its level of activity during the early part of the season (Priestley, 1930; Friesner, 1941; Pearson, 1950; Fraser, 1956; Bannan, 1955; Eggler, 1955; Glock, 1955; Wareing, 1958; Kozlowski *et al.*, 1962; Reinders-Gouwentak, 1965). These effects are partly indirect; cambial swelling depends on water uptake, and thus on soil temperature, while cambial division is stimulated by bud reactivation. Bannan found that cambial swelling occurs as soon as the frost leaves the ground near Toronto, Canada, and Fraser that the time of initiation of cambial activity varies from year to year according to spring temperatures at Chalk River, Ontario. In Louisiana in the southern United States Eggler found that cambial activity begins sooner in willows growing in poorly drained swamp than in those growing in the well-drained batture which is adjacent to the river, and concluded that this is probably because air temperatures are lower in the batture due to the cooling effect of river water. Daubenmire (1949) found that cambial activity is not initiated in *Robinia pseudoacacia* when night temperatures are below $1 \cdot 1 °C$, although there was no correlation with temperature in other species studied. During the growing season cambial activity and temperature may show a positive correlation if water is not limiting (Ladefoged, 1952; Glock, 1955; Fritts, 1958). Decrease in ring width with increasing altitude is probably due to the direct and indirect effects of decreased temperatures on cambial activity (Wareing, 1958). Periods of low temperature during the growing season may bring about the slowing or cessation of radial growth (Daubenmire and Deters, 1947). Handley (1939) subjected the woody shoots of young saplings of *Fraxinus excelsior* L. and *Acer pseudoplatanus* to a continuous temperature of $2°C$ during the normal growing season, with the result that radial growth was almost completely inhibited throughout the woody stem, although terminal growth was normal except that it started later and proceeded more slowly. High temperatures sometimes show an inverse correlation with radial growth because they increase the transpiration rate. This may have the effect of depressing terminal and cambial growth if there is a shortage of available water (Friesner, 1941; Fritts, 1958). Fritts has shown that high temperatures would not reduce growth rates if humidity remained constant. The cessation of cambial growth may be brought about by temperature, although day length and shortage of available water are apparently more common causes. Waisel and Fahn (1965) found that cambial growth ceases only under a combination of low temperatures and short days in *Robinia pseudoacacia*, suggesting an interaction between temperature and photoperiod.

In the non-dormant state the cambium is susceptible to frosts, which may merely decrease the growth rate or may injure or kill the cambial cells. Frost injury leads to the production of abnormal xylem cells which are discernible as dark spots or circles in transverse sections of the wood (Fig. 9.7), and which are known as frost spots and frost rings (Glock, 1951; Ladefoged, 1952; Glock and Agerter, 1962; Parker, 1963; Reinders-Gouwentak, 1965; Glerum and Farrar, 1966). Some trees may be adapted to withstand frosts during the growing season; Daubenmire and Deters (1947) found no correlation between the frost-free season and the season of radial growth, and Fraser (1956) found that radial growth slowed down independently of autumn frosts.

Light intensity is an important factor in radial growth when it affects the rate of photosynthesis. It is well known that trees in close stands tend to have narrow rings and that radial growth is much greater in dominant than in suppressed trees, although this effect may be brought about by competition for water and nutrients as well as light. However, Pearson (1950) carried out experiments to determine the effects of shade and found that it leads to the development of slender, cylindrical boles due to a reduction in radial growth, which is greater towards the base of the tree. Trees with adequate light develop wider, tapered boles. Fritts (1958) found a correlation between radial growth and percentage of possible sunshine on the preceding day and considered that this was probably due to the effect of light on photosynthesis and the delay between the production of food and its arrival in the cambium of the trunk.

Photoperiod affects the cambium indirectly through the buds or the mature leaves. It sometimes determines the time of cessation of cambial activity when temperature or available moisture are not limiting. Eggler (1955), studying radial growth in nine species of trees in southern Louisiana, found that cambial activity ceased at the same time in all ten individuals of one species, suggesting photoperiodic control, while in other species the time of cessation varied. Some species ceased activity very early in spite of adequate moisture and temperature, but in one of these cambial activity both began and ceased under lengthening days. Downs and Borthwick (1956) found that seedlings of some species could be kept growing continuously under day lengths of 16 hr, while those of other species grew intermittently or became dormant. It is apparent that the cessation of cambial activity may be controlled by factors other than temperature, available moisture, and photoperiod.

In many woody species daylength affects the duration of terminal growth, with long days promoting and short days bringing about the cessation of growth. When cambial growth is dependent on terminal growth, as in many diffuse-porous species, cessation of terminal growth under shortening days will bring about the cessation of cambial activity. In conifers and ring-porous species, in which cambial activity continues after the cessation of terminal growth, daylength may again determine the time at which the cambium becomes dormant (Wareing, 1958). Continued cambial activity in these species is stimulated by the production of auxin in the mature leaves, and this production may be influenced by photoperiod. In *Pinus sylvestris*, Wareing (1951a) has shown that in the absence of bud growth cambial activity is prolonged under long days and stopped under short days. In *Robinia pseudoacacia* the situation is less clear. Mollart (1954) found that in the absence of bud growth cambial activity ceased under short days and continued in a high proportion of the plants grown under long days. Wareing and Roberts (1956) obtained similar results, with cambial activity being maintained in about half the plants growing under long days. Waisel and Fahn (1965) found that in this species photoperiod affected terminal growth and the type of wood produced, rather than cambial activity. Radial growth continued under both long and short days, provided that temperatures were high; with low temperatures it ceased under short days and continued very slowly under long days. It appears that cessation of cambial activity in *Robinia* may be controlled by an interaction of photoperiod, temperature, and auxin production.

Pauley (1958) discusses photoperiod in relation to the cessation of growth and points out that in areas with alternate frost and frost-free seasons the survival of perennials depends on a regular, dependable timing mechanism to stop growth early enough to avoid excessive frost damage. The only environmental factor which is highly regular from year to year is daylength. Photoperiodic ecotypes are known to exist (Pauley and Perry, 1954; Vaartaja, 1954; Wassink and Wiersma, 1955), and Moshkov (1933, 1934, 1935) found that southern tree species grown at Leningrad in Russia, where the maximum daylength is 20 hr, continued to grow until killed by frost instead of becoming dormant at the end of the summer.

Competition may depress the radial growth rate of trees through its effect on the crown and roots. Kozlowski and Peterson (1962) found that suppressed red pine trees grow more slowly, more erratically, and for a shorter period than dominant ones. Competition for light, water, and minerals affects the ability of the tree to produce carbohydrates and hormones, and root growth is inhibited by hormone and food deficiency, causing decreased water and mineral uptake. Larson (1962) discusses the relationship between cambial activity and auxin in long-crowned, open-grown trees and small-crowned trees growing under more restricted conditions. Cambial activity has been reported to begin earlier, spread more rapidly down the tree, and then proceed more vigorously in long-crowned trees. Auxin is produced in greater amounts in long-crowned trees, since there are more branches, and the branches extend farther down the bole, facilitating the spread of auxin to the base of the trunk. The cambium is as active at the base of the bole as at the top, and a tapered stem is produced. In small-crowned trees there may be a lack of auxin at the base of the bole, resulting in the formation of narrow rings and a cylindrical stem form. Release from competition usually acts as an indirect stimulus for radial growth through the increased availability of light and soil moisture. Although light has usually been regarded as the most important factor in release from competition, soil moisture may be more important in dry climates (Pearson, 1950). Because of the close relationship between terminal and radial growth, there is usually a positive correlation between crown size and cambial activity. An exception occurs in ponderosa pine in the south-eastern United States, in which photosynthesis is limited by available water rather than crown size (Pearson, 1950).

Oxygen and carbon dioxide levels, wind and defoliating agents may affect radial growth. High carbon dioxide levels in the soil have been shown to retard root activity, although the decreased activity of roots in flooded or poorly drained soils is probably due in most species to lack of oxygen rather than toxic levels of carbon dioxide (Voigt, 1962). Atmospheric carbon dioxide may indirectly affect cambial activity if it is the limiting factor in photosynthesis. The photosynthetic rate may be increased by low-velocity winds which improve the carbon dioxide supply. It may be decreased by the effect of wind in increasing the transpiration rate and causing a water deficit in the leaves. The stomata may partially close, reducing the intake of carbon dioxide (Satoo, 1962). Strong winds often cause asymmetry in plant form, with greater growth on the leeward side. This is accompanied by stronger cambial activity on the leeward side than on the windward one, where it may cease altogether (Oosting, 1956, p. 144). Jacobs (1954) has studied the effect of wind sway on radial growth in *Pinus radiata*. Cambial activity in the lower trunk is greater in trees which sway than in those which are held rigid, and after ten years the effect is detectable at heights of up to 30 ft. Sway tends to cause eccentric trunk development, with greater cambial activity on the leeward side. Defoliation usually brings about a reduction in radial growth (Glock, 1955; Mott *et al.*, 1957; Studhalter *et al.*, 1963) due to reduced carbohydrate and auxin production. Increases in cambial activity following defoliation have been reported (Glock, 1955); in such cases radial growth may increase, at least temporarily, because of decreased competition from the buds and leaves for water.

Cambial activity is indirectly influenced by factors such as topography, altitude, and latitude which modify environmental conditions. Topography modifies rainfall, temperature, and light intensity, gives rise to differences in soil properties, and affects run-off, drainage, and erosion. In the Northern Hemisphere south-facing slopes receive more light, have higher temperatures, and are drier than average sites. Latitude affects light intensity, photoperiod, and temperature, while altitude modifies light intensity, temperature, and moisture. The period of cambial activity may be shifted in time and reduced in length, and rings may become narrower with increasing altitude. In dry regions drought increases in importance at lower altitudes and temperature at higher altitudes; with increasing height the onset of cambial activity may be delayed by low temperatures, and the effects of dry summer weather may be apparent later (Fowells, 1941; Daubenmire, 1945; Lobzhanidze, 1963).

Cambial activity may be affected by pathogens; Pfeiffer (1926), Boyce (1948), and Butler and Jones (1949) have reviewed the morphological responses of plants to the presence of pathogens. The stimulation of cambial activity by insects, fungi, bacteria, and mistletoes results in the formation of galls or tumours. The pattern of division in the cambium may be affected, with disorientation of the planes of division or altered proportions of fusiform to ray cells. The cambial derivatives are often abnormal, and most commonly take the form of lignified parenchyma. Cambial activity beyond the point of infection may in some cases be reduced, probably because of shortage of carbohydrates or other growth materials. The effect of the pathogen on the general health of the plant may be reflected in decreasing amounts of radial growth. White and Millington (1954) have described a woody tumour in *Picea glauca* which is probably caused by the crown gall bacterium or a similar agent. It causes an increase in the amount of cambial activity, partly due to extension of the growing season. Most extra cells are produced towards the xylem, and phloem production is at most only slightly affected. The tumour arises from one or a few cells which are usually infected during the first year's growth; infected cambial cells increase in number by anticlinal division, and this together with cell elongation increases the size of the tumour. The planes of division of the cambial cells may be irregular, and this is reflected in an irregular arrangement of their derivatives.

The effects of balsam woolly aphid infestation on cambial activity in *Abies grandis* (Dougl.) Lind. have been described by Smith (1967). Salivary secretions injected into the cortex or outer phloem induce the production of wood that is in some respects similar to compression wood. Cambial growth after infestation is characterized by increased rates of periclinal and anticlinal division of fusiform initials, increased production of new ray initials from fusiform initials and from anticlinal divisions in existing ray initials, and decline of numerous fusiform initials and termination of many files by maturation. Extensive loss of fusiform initials results in increased frequency of ray fusion and increased frequency of ray splitting by the intrusion of elongating fusiform initials; both the size and the number of rays are markedly increased. Menzies (1964) has described the effects of a mistletoe (*Loranthus micranthus* Hook. f.). The haustorium of the parasite stimulates cambial activity in the host, and the production of secondary tissue is increased for some distance from the infection. In the immediate vicinity of the haustorium there is disorientation of the planes of cambial division and the rate of division is accelerated. A gall is produced which may be four or five times the diameter of an uninfected stem and is made up of mainly lignified parenchyma, with the cells orientated in all directions. Dwarf mistletoe (*Arceuthobium*) was found by Srivastava and Esau (1961) to affect mainly the rays of host conifers. Extensive ray fusions occur due to loss of intervening fusiform initials, subdivision of fusiform initials, intrusive growth of marginal ray cells, and the incorporation of parasite sinkers into the rays. Infection may result in an increase in the number of rays, but no significant increase in ring width that could be definitely related to the parasitic attack was found. The root parasite *Exocarpus bidwillii* Hook. f. (Santalaceae) is unusual in that it seldom causes abnormal behaviour in the cambium of the host. The xylem surrounding the sucker is apparently quite normal, although sometimes more xylem is formed on the side where the branches are embedded (Fineran, 1963). In the portion of the root distal to the infection cambial activity is sometimes markedly retarded, probably because of a shortage of carbohydrates.

REFERENCES

ALFIERI, F. J. and EVERT, R. F. (1968). Seasonal development of the secondary phloem in *Pinus*. *Am. J. Botany* **55**, 518–28.

ALVIM, P. DE T. (1964). Tree growth and periodicity in tropical climates. In ZIMMERMANN, M. H. (ed.), *The Formation of Wood in Forest Trees*, pp. 479–95. Academic Press, New York and London.

ANTEVS, E. (1925). The big tree as a climatic measure. *Carnegie Inst. Wash., Publ.* **352**, 115–53.

ARTSCHWAGER, E. (1945). Growth studies on guayule (*Parthenium argentatum*). *U.S. Dept. Agr. Tech. Bull.* **885**, 1–19.

— (1950). The time factor in the differentiation of secondary xylem and phloem in pecan. *Am. J. Botany* **37**, 15–24.

AVERY, G. S. JR., BURKHOLDER, P. R. and CREIGHTON, H. B. (1937). Production and distribution of growth hormone in shoots of *Aesculus* and *Malus*, and its probable role in stimulating cambial activity. *Am. J. Botany* **24**, 51–8.

BAILEY, I. W. (1943). Some misleading terminologies in the literature of 'plant tissue culture'. *Science, N.Y.* **98**, 539.

— (1944). The comparative anatomy of the Winteraceae. III. Wood. *J. Arnold Arbor.* **25**, 97–103.

BANNAN, M. W. (1950). The frequency of anticlinal divisions in fusiform cambial cells of *Chamaecyparis*. *Am. J. Botany* **37**, 511–19.

— (1951). The annual cycle of size changes in the fusiform cambial cells of *Chamaecyparis* and *Thuja*. *Can. J. Botany* **29**, 421–37.

— (1955). The vascular cambium and radial growth in *Thuja occidentalis* L. *Can. J. Botany* **33**, 113–38.

— (1960). Ontogenetic trends in conifer cambium with respect to frequency of anticlinal division and cell length. *Can. J. Botany* **38**, 795–802.

— (1962). The vascular cambium and tree-ring development. In KOZLOWSKI, T. T. (ed.), *Tree Growth*, pp. 3–21, Ronald Press, New York.

— (1964). Tracheid size and anticlinal divisions in the cambium of *Pseudotsuga*. *Can. J. Botany* **42**, 603–31.

BOYCE, J. S. (1948). *Forest Pathology* (*2nd ed.*), McGraw Hill, New York, Toronto, London.

BROWN, H. P. (1915). Growth studies in forest trees. II. *Pinus strobus* L. *Botan. Gaz.* **59**, 197–241.

BUTLER, E. J. and JONES, S. G. (1949). *Plant Pathology*, Macmillan, London.

CATESSON, A. M. (1964). Origine, fonctionnement et variations cytologiques saisonnières du cambium de l'*Acer pseudoplatanus* L. (Acéracées). *Ann. Sci. nat. (Bot.) 12e Sér.* **5**, 229–498.

CHOWDHURY, K. A. (1958). Extension and radial growth in tropical perennial plants. *Proc. Delhi Univ. Seminar* 1957, 138–9.

— (1961). Growth rings in tropical trees and taxonomy. *Pac. Sci. Congr. 10th, Abstracts*, 280.

— and TANDAN, K. N. (1950). Extension and radial growth in trees. *Nature, Lond.* **165**, 732–3.

CUMBIE, B. G. (1967). Developmental changes in the vascular cambium in *Leitneria floridana. Am. J. Botany* **54**, 414–24.

DAUBENMIRE, R. F. (1945). Radial growth of trees at different altitudes. *Botan. Gaz.* **107**, 463–7.

— (1949). Relation of temperature and daylength to the inception of tree growth in spring. *Botan. Gaz.* **110**, 464–75.

— and DETERS, M. E. (1947). Comparative studies of growth in deciduous and evergreen trees. *Botan. Gaz.* **109**, 1–12.

DAVIS, J. D. (1968). Seasonal cycle of phloem development in *Parthenocissus inserta. Am. J. Botany* **55**, 716 (Abstr.).

— and EVERT, R. F. (1966). Phloem development in *Celastrus scandens. Am. J. Botany* **53**, 616 (Abstr.).

— and — (1968). Seasonal development in the secondary phloem in *Populus tremuloides. Botan. Gaz.* **129**, 1–8.

DERR, W. F. and EVERT, R. F. (1967). The cambium and seasonal development of the phloem in *Robinia pseudoacacia. Am. J. Botany* **54**, 147–53.

DIGBY, J. and WAREING, P. F. (1966a). The effect of applied growth hormones on cambial division and the differentiation of the cambial derivatives. *Ann. Botany* **30**, 539–48.

— and — (1966b). The relationship between endogenous hormone levels in the plant and seasonal aspects of cambial activity. *Ann. Bot., N.S.* **30**, 607–22.

DOWNS, R. J. and BORTHWICK, H. A. (1956). Effects of photoperiod on the growth of trees. *Botan. Gaz.* **117**, 310–26.

EGGLER, W. A. (1955). Radial growth in nine species of trees in southern Louisiana. *Ecol.* **36**, 130–6.

ESAU, K. (1948). Phloem structure in the grapevine, and its seasonal changes. *Hilgardia* **18**, 217–96.

EVERT, R. F. (1960). Phloem structure in *Pyrus communis* L. and its seasonal changes. *Univ. Calif. Publ. Bot.* **32**, 127–94.

— (1961). Some aspects of cambial development in *Pyrus communis. Am. J. Botany* **48**, 479–88.

— (1962). Some aspects of phloem development in *Tilia americana. Am. J. Botany* **49**, 659 (Abstr.).

— (1963). The cambium and seasonal development of the phloem in *Pyrus malus. Am. J. Botany* **50**, 149–59.

FAHN, A. (1953). Annual wood ring development in maquis trees of Israel. *Palest. J. Bot.* **6**, 1–26.

— (1955). The development of the growth ring in wood of *Quercus infectoria* and *Pistacia lentiscus* in the hill region of Israel. *Trop. Woods* **101**, 52–9.

— (1959). Xylem structure and annual rhythm of development in trees and shrubs of the desert. II. *Acacia tortilis* and *A. raddiana. Bull. Res. Counc. Israel* **7D**, 23–8.

FINERAN, B. A. (1964). Studies on the root parasitism of *Exocarpus bidwillii* Hook f. IV. Structure of the mature haustorium. *Phytomorphol.* **13**, 249–67.

FOWELLS, H. A. (1941). The period of seasonal growth of ponderosa pine and associated species. *J. Forest.* **39**, 601–8.

FRASER, D. A. (1956). Ecological studies of forest trees at Chalk River, Ontario, Canada. II. Ecological conditions and radial increment. *Ecol.* **37**, 777–89.

— (1962). Tree growth in relation to soil moisture. In KOZLOWSKI, T. T. (ed.), *Tree Growth*, pp. 183–204, Ronald Press, New York.

FRIESNER, R. C. (1941). A preliminary study of growth in the beech, *Fagus grandifolia,* by the dendrograph method. *Butler Univ. Bot. Stud.* **5**, 85–94.

FRITTS, H. C. (1958). An analysis of radial growth of beech in a central Ohio forest during 1954–1955. *Ecol.* **39**, 705–20.

GILL, N. (1932). The phloem of ash (*Fraxinus excelsior* Linn.). Its differentiation and seasonal variation. *Proc. Leeds Phil. Lit. Soc., Sci. Sect.* **2**, 347–55.

— (1933). The relation of flowering and cambial activity. Observations on vascular differentiation and dry-weight changes in the catkins of some early flowering catkin-bearing dicotyledons. *New Phytol.* **32**, 1–12.

GLERUM, C. and FARRAR, J. L. (1966). Frost ring formation in the stems of some coniferous species. *Can. J. Botany* **44**, 879–86.

GLOCK, W. S. (1941). Growth rings and climate. *Botan. Rev.* **7**, 649–713.

— (1951). Cambial frost injuries and multiple growth layers at Lubbock, Texas. *Ecol.* **32**, 28–36.

— (1955). Tree growth. II. Growth rings and climate. *Botan. Rev.* **21**, 73–188.

— and AGERTER, S. R. (1962). Rainfall and tree growth. In KOZLOWSKI, T. T. (ed.), *Tree Growth*, pp. 23–53, Ronald Press, New York.

GRILLOS, S. J. and SMITH, F. H. (1959). The secondary phloem of Douglas-fir. *Forest Sci.* **5**, 377–88.

GROSSENBACHER, J. G. (1915). The periodicity and distribution of radial growth in trees and their relation to the development of 'annual rings'. *Trans. Wis. Acad. Sci.* **18**, 1–77.

HANDLEY, W. R. C. (1939). The effect of prolonged chilling on water movement and radial growth in trees. *Ann. Bot., N.S.* **3**, 803–13.

HEINECKE, A. J. (1937). Some cultural conditions influencing the manufacture of carbohydrates by apple leaves. *N.Y. Hort. Soc. Proc.* 149–56.

HOLTERMANN, C. (1902). Anatomisch-physiologische Untersuchen in den Tropen. *Sitzb. Kön. Preuss. Akad. Wiss. Berlin,* 1902 **(1),** 656–74. Cited after STUDHALTER (1955).

HUBER, B. and JAZEWITSCH, W. V. (1956). Tree-ring studies of the Forestry-Botany Institute of Tharandt and Munich. *Tree-Ring Bull.* **21**, 28–30.

JACOBS, M. R. (1954). The effect of wind sway on the form and development of *Pinus radiata. Australian J. Botany* **2**, 35–51.

KENNEDY, R. W. and FARRAR, J. L. (1965). Tracheid development in tilted seedlings. In CÔTÉ, W. A. (ed.), *Cellular Ultrastructure of Woody Plants*, pp. 419–53, Syracuse University Press, New York.

KIENHOLZ, R. (1934). Leader, needle, cambial and root growth of certain conifers and their interrelations. *Botan. Gaz.* **96**, 73–92.

KNUDSON, L. (1913). Observations on the inception, season, and duration of cambium development in the American larch (*Larix laricina* (Du Roi) Koch). *Bull. Torrey Bot. Club,* **40**, 271–93.

KORIBA, K. (1958). On the periodicity of tree-growth in the tropics, with reference to the mode of branching, the leaf fall, and the formation of the resting bud. *Gard. Bull. Straits Settlements* **17**, 11–81.

KOZLOWSKI, T. T. (1962). Photosynthesis, climate, and tree growth. In KOZLOWSKI, T. T. (ed.), *Tree Growth,* pp. 149–64, Ronald Press, New York.

— and PETERSON, T. A. (1962). Seasonal growth of dominant, intermediate and suppressed red pine trees. *Botan. Gaz.* **124**, 146–54.

—, WINGET, C. H. and TORRIE, J. H. (1962). Daily radial growth of oak in relation to maximum and minimum temperature. *Botan. Gaz.* **124**, 9–17.

KRAMER, P. J. (1962). The role of water in tree growth. In KOZLOWSKI, T. T. (ed.), *Tree Growth,* pp. 171–82, Ronald Press, New York.

— and COILE, T. S. (1940). An estimation of the volume of water made available by root extension. *Plant Physiol.* **15**, 743–7.

LADEFOGED, K. (1952). The periodicity of wood formation. *Kgl. Danske Videnskab Selskab. Biol. Skrifter* **7**, 1–98.

LARSON, P. R. (1962). Auxin gradients and the regulation of cambial activity. In KOZLOWSKI, T. T. (ed.), *Tree Growth,* pp. 97–117, Ronald Press, New York.

LOBZHANIDZE, E. D. (1966). The influence of vertical zoning on cambial activity. Abstract in *Biol. Abstr.* **47**, 34104.

MENZIES, B. P. (1954). Seedling development and haustorial system of *Loranthus micranthus* Hook. f. *Phytomorphol.* **4**, 397–409.

MIKOLA, P. (1962). Temperature and tree growth near the northern timber line. In KOZLOWSKI, T. T. (ed.), *Tree Growth,* pp. 265–74. Ronald Press, New York.

MOLLART, D. L. (1954). The control of cambial activity in *Robinia pseudoacacia.* M.Sc. Thesis, Univ. of Manchester, cited after WAREING (1956).

MORRIS, R. F. (1951). The effects of flowering on the foliage production and growth of balsam fir. *Forest. Chron.* **27**, 40–57.

MOSHKOV, B. S. (1933). Photoperiodicity of certain woody species. Abstract in *Biol. Abstr.* **7**, 20678.

— (1934). Photoperiodicity of trees and its practical importance. Abstract in *Biol. Abstr.* **8**, 1680.

— (1935). Photoperiodismus und Frostharte aus dauernder Gewachse. *Planta* **23**, 774–803.

MOTT, D. G., NAIRN, L. D. and COOK, J. A. (1957). Radial growth in forest trees and effects of insect defoliation. *Forest Sci.* **3**, 286–304.

OOSTING, H. J. (1956). *The Study of Plant Communities,* 2nd ed., Freeman, San Francisco.

PARKER, J. (1963). Cold resistance in woody plants. *Botan. Rev.* **29**, 123–201.

PAULEY, S. S. (1958). Photoperiodism in relation to tree improvement. In THIMANN, K. V. (ed.), *The Physiology of Forest Trees,* pp. 557–71, Ronald Press, New York.

— and PERRY, T. O. (1954). Ecotypic variations of the photoperiodic response in *Populus. J. Arnold Arbor.* **35**, 167–88.

PEARSON, G. A. (1950). Management of ponderosa pine in the Southwest. *U.S. Dept. Agr., For. Ser., Agr. Mon.* **6**, 1–218.

PFEIFFER, H. (1926). Das abnorme Dickenwachstum. In LINSBAUER, K., *Handbuch der Pflanzenanatomie*, Abt. II, Teil 2, Bd. IX, Bornträger, Berlin.

PRIESTLEY, J. H. (1930). Studies in the physiology of cambial activity. III. The seasonal activity of the cambium. *New Phytol.* **29**, 316–54.

RAATZ, W. (1892). Die Stabbildungen im secundären Holzkörper der Bäume und die Initialentheorie. *Jahrb. wiss. Bot.* **23**, 567–636.

REED, H. S. and MACDOUGAL, D. T. (1937). Periodicity in the growth of the orange tree. *Growth* **1**, 371–3.

REINDERS-GOUWENTAK, C. A. (1965). Physiology of the cambium and other secondary meristems of the shoot. In RUHLAND, W. (ed.), *Encyclopaedia of Plant Physiology XV*, **(1)**, 1077–105, Springer, Berlin.

— and VAN DER VEEN, J. H. (1953). Cambial activity in *Populus* in connection with flowering and growth hormone. *Proc. kon. ned. Akad. Wet.* **C56**, 194–201. Cited after REINDERS-GOUWENTAK (1965).

SATOO, T. (1962). Wind, transpiration and tree growth. In KOZLOWSKI, T. T. (ed.), *Tree Growth*, pp. 299–310, Ronald Press, New York.

SMITH, F. H. (1967). Effects of balsam woolly aphid (*Adelges piceae*) infestation on cambial activity in *Abies grandis*. *Am. J. Botany* **54**, 1215–23.

SRIVASTAVA, L. M. and ESAU, K. (1961). Relations of the dwarf-mistletoe (*Arceuthobium*) to the xylem tissue of conifers. II. Effect of the parasite on the xylem anatomy of the host. *Am. J. Botany* **48**, 209–15.

STUDHALTER, R. A. (1955). Tree growth. I. Some historical chapters. *Botan. Rev.* **21**, 1–72.

—, GLOCK, W. S. and AGERTER, S. R. (1963). Tree growth. *Botan. Rev.* **29**, 245–365.

TEPPER, H. B. and HOLLIS, C. A. (1967). Mitotic reactivation of the terminal bud and cambium of white ash. *Science, N.Y.* **156**, 1635–6.

TINGLEY, M. A. (1936). Double growth rings in red astrakan. *Proc. Am. Soc. Hort. Sci.* **34**, 61.

TUCKER, C. M. (1968). Seasonal phloem development in *Ulmus americana*. *Am. J. Botany* **55**, 716 (Abstr.).

— and EVERT, R. F. (1969). Seasonal development of the secondary phloem in *Acer negundo*. *Am. J. Botany* **56**, 275–84.

VAARTAJA, O. (1954). Photoperiodic ecotypes of trees. *Can. J. Botany* **32**, 392–9.

VOIGT, G. K. (1962). The role of carbon dioxide in soil. In KOZLOWSKI, T. T. (ed.), *Tree Growth*, pp. 205–20, Ronald Press, New York.

WAISEL, Y. and FAHN, A. (1965). The effects of environment on wood formation and cambial activity in *Robinia pseudoacacia* L. *New Phytol.* **64**, 436–42.

WALTER, H. (1963). The water supply of desert plants. In RUTTER, A. J. and WHITEHEAD, F. H. (eds.), *The Water Relations of Plants*, Blackwell, London.

WAREING, P. F. (1951a). Growth studies in woody species. III. Further photoperiodic effects in *Pinus sylvestris*. *Physiol. Plant.* **4**, 41–56.

—(1951b). Growth studies in woody species. IV. Initiation of cambial activity in ring-porous species. *Physiol. Plant.* **4**, 546–62.

— (1956). Photoperiodism in woody plants. *Ann. Rev. Plant Physiol.* **7**, 191–214.

— (1958). The physiology of cambial activity. *J. Inst. Wood Sci.* **1**, 34–42.

WAREING, P. F. and BLACK, M. (1958). Photoperiodism in seeds and seedlings of woody species. In THIMANN, K. V. (ed.), *The Physiology of Forest Trees*, pp. 529–56, Ronald Press, New York.

— and ROBERTS, D. L. (1956). Photoperiodic control of cambial activity in *Robinia pseudoacacia*. *New Phytol.* **55,** 356–66.

WASSINK, E. C. and WIERSMA, J. H. (1955). Daylength responses of some forest trees. *Acta Bot. Neerland.* **4,** 657–70.

WHITE, P. R. and MILLINGTON, W. F. (1954). The structure and development of a woody tumor affecting *Picea glauca*. *Am. J. Botany* **41,** 353–61.

WHITMORE, F. W. and ZAHNER, R. (1966). Development of the xylem ring in stems of young red pine trees. *Forest Sci.* **12,** 198–210.

WIGHT, W. (1933). Radial growth of the xylem and the starch reserves of *Pinus sylvestris*: A preliminary survey. *New Phytol.* **32,** 77–96.

WILCOX, H. (1962). Cambial growth characteristics. In KOZLOWSKI, T. T. (ed.), *Tree Growth*, pp. 57–88, Ronald Press, New York.

WILSON, B. F. (1964). A model for cell production by the cambium of conifers. In ZIMMERMAN, M. H. (ed.), *The Formation of Wood in Forest Trees*, pp. 19–36, Academic Press, New York and London.

— (1966). Mitotic activity in the cambial zone of *Pinus strobus*. *Am. J. Botany* **53,** 364–72.

WORT, D. J. (1962). Physiology of cambial activity. In KOZLOWSKI, T. T. (ed.), *Tree Growth*, 89–95, Ronald Press, New York.

ZIMMERMAN, M. H. (ed.) (1964). *The Formation of Wood in Forest Trees*, Academic Press, New York and London.

10

Experimental control of cambial development

The close correspondence between the activity of the cambium, which has been considered in the previous chapter, and the resumption of growth by the buds is well known. Experimental disbudding is followed by a failure of cambial activity. This applies even in ring-porous trees, in which cambial activity may begin before the buds swell and spreads rapidly throughout the tree. It is true that such trees may form slight amounts of spring wood when disbudded, but Wareing (1951) has shown that this is due to the development of adventitious buds on the wounds caused by the disbudding. In diffuse-porous trees it appears that this slight amount of bud growth is insufficient to stimulate the cambial initials to divide. This relationship between bud growth and cambial activity no doubt also applies to conifers. At least in *Pinus* disbudding resulted in only slight activity which produced abnormal xylem derivatives (Jost, 1893).

It is natural to conclude that the basipetal spread of activity below expanding buds is due to the downward movement of a stimulus. Experiments in which rings of bark are removed from shoots support this hypothesis. If these rings are deep enough to include the cambium the downward progression of the resumption of divisions in the cambium halts at the ring, the cambium below it remaining inactive, or nearly so (Brown, 1936).

In early experiments by Snow (1933) it was demonstrated that the stimulus which activates the cambium could pass from one plant to another if cut surfaces were brought into contact, and could even be transmitted through a barrier of moist muslin. Finally, Snow (1935) obtained direct proof of the hormonal nature of the stimulus by demonstrating a striking rise in cambial activity in *Helianthus* seedlings on application of natural hormone. Subsequent workers have shown that synthetic indole-3-acetic acid can replace expanding buds in inducing cambial activity and that the activity may be enhanced by a variety of auxins, gibberellins, and vitamins (Reinders-Gouwentak, 1965). However, Söding (1940) showed that hormones derived from shoots of *Acer* were more effective in inducing activity in *Helianthus* stems than a comparable amount of IAA. The cambium is evidently influenced by a complex of endogenous growth hormones. The particular roles played by some of these will be discussed presently.

When the distribution of auxin within the tree is investigated by bioassay it is found to follow a pattern in space and time which corresponds with that of cambial activity. There is a clear build-up of auxin concentration in spring to high levels during the period of maximum activity, and this increase in concentration first occurs high in the stem and later at a lower level (Shepherd and Rowan, 1966). In the dormant condition the presence of auxin cannot be demonstrated (Söding, 1940). Auxin levels fall off rapidly in diffuse-porous species at a time when shoot elongation ceases and xylem formation is completed (Avery *et al.*, 1937). In ring-porous trees, on the other hand, xylem formation continues after elongation growth has finished. With long days, low but sufficient levels of auxin are maintained in the cambium under the influence of the mature leaves present on the shoot (Wareing and Roberts, 1956). Not only the onset of dormancy is different in these two classes of trees but also, as will be seen presently, their two tissues, xylem and phloem, behave differently in relation to auxin levels and the duration of activity.

Auxin moves down the tree at comparable rates to the spread of cambial activity both in ring-porous trees (rapid spread) and diffuse-porous trees (gradual spread). The speed at which the increase in auxin concentration travels down the trunk of ring-porous trees led to the suggestion that a precursor of auxin is present in the cambium before bud-break and that it is a stimulus which triggers off its transformation into auxin which is transmitted from the buds, rather than the hormone itself. This hypothesis is supported by Digby and Wareing (1966b), who found that in *Ulmus glabra* Huds. a substance was present in the cambium throughout the tree which promoted growth of *Avena* coleoptiles and which chromatographed with the Pf of tryptophane. As soon as buds had swollen, extracts from the cambium showed that all down the trunk this compound had been replaced by IAA. Tryptophane itself, or a similar compound, could well be a precursor of IAA.

The cambium will not respond to the presence of auxin at all times. When the cambium is dormant applications of auxin have been found to induce activity only for a few centimetres below the cut surface. This dormancy can be broken artificially by exposure to the vapour of ethylene chlorohydrin (Reinders-Gouwentak, 1949).

The different roles played by IAA and GA in activating the cambium were studied by Wareing *et al.* (1964) using coniferous and dicotyledonous trees. GA and IAA applied independently both promoted cambial activity, but only auxin appeared to induce the differentiation of the xylem derivatives. Morey and Cronshaw (1968) also found GA to stimulate cambial activity locally in seedlings of *Acer rubrum* in regions where the endogenous rate of cambial activity was low. Like Wareing *et al.*, Morey and Cronshaw also found GA alone was without effect on the development of the secondary xylem. Nevertheless, they draw attention to reports that in conifers and herbaceous dicotyledons GA may affect xylem differentiation. Okunda (1959), for example, found that GA accelerates differentiation of cambial derivatives in *Pharbitis nil* without stimulating division.

In a study by Digby and Wareing (1966a) gibberellic acid and indole-3-acetic acid were applied to non-dormant disbudded shoots of *Populus robusta* Schneid. It was found that not only the absolute levels of these hormones were important but also the balance between them. IAA applied alone stimulated cambial divisions, and the derivatives differentiated to produce xylem tissue. No phloem was formed. GA by itself also induced divisions, but the few derivatives to the inner side remained undifferentiated. Considerable amounts of phloem were formed. De Maggio (1966) also induced phloem differentiation by application of GA to explants from dormant shoots of *Pinus strobus*. Application of both hormones together gave greater production. Fully formed xylem was at a maximum at high concentrations of IAA combined with low concentrations of GA (in p.p.m. 500 IAA/100 GA). Conversely, maximum phloem production occurred with low levels of IAA and high levels of GA (in p.p.m. 100 IAA/500 GA).

A range of concentrations of IAA (from 1 to 1,000 p.p.m., with GA at 100 p.p.m. in all cases) was applied to cut shoots of the ring-porous species *Robinia pseudoacacia*. An increase in vessel diameter occurred with increasing IAA concentration. With low levels of IAA, the xylem elements were mainly tracheids and rectangular in transverse section. The tissue was therefore similar to the late wood which forms after shoot elongation has ceased, at which time, as we have seen, IAA levels are low in the shoots. On the other hand, with high levels of IAA the radial diameters of the xylem elements are much greater and the pores tend to be more circular. This xylem resembles the early wood produced during the period of active shoot elongation when IAA content of the shoot is high (Digby and Wareing, 1966b). It thus seems justifiable to conclude that the differences between early and late wood in ring-porous trees are due to differing seasonal levels of auxin reaching the cambium. These results are also in accord with observations that phloem production and xylem production do not extend over the same periods (see preceding chapter).

Attempts have been made to align these effects of hormones on xylem production and differentiation with the promotion of elongation of cells in etiolated tissues, such as the classical example of the oat coleoptile. It is generally held that, if high levels of auxin are present as cambial derivatives are enlarging radially, as in the early growing season, then xylem elements of wide diameter will be promoted. Since cambial derivatives increase in diameter too slowly for direct measurement of IAA-induced expansion, Whitmore (1968) employed the incorporation of labelled sugar into cell-wall carbohydrates as a measure of primary wall expansion. Cambial derivatives of *Populus* were found to respond as expected under the hypothesis of IAA-induced cell-expansion. However, this was not so in *Pinus*, even though it was shown that IAA was not destroyed, could enter *Pinus* cells, and was not in a bound state. Whitmore concluded that there were real differences between the auxin-cell wall relations present in *Populus* and *Pinus*, and that no single theory of auxin control of radial expansion of cambial derivatives can be applied to both gymnosperms and angiosperms.

The supply of water to a tree is clearly a decisive factor in growth. More specifically the rate of cambial division in trees has been shown to be intimately correlated with the moisture content of the soil (Shepherd, 1964; Whitmore and Fahner, 1966). The relationship between water potential and cambial development can be studied more directly by using cultures of detached portions of stems. In this way living cambial cells can be maintained in continuous contact with osmotic solutions. Using this approach, Doley and Leyton (1968) combined various concentrations of IAA and GA with various conditions of controlled water potential, and were able to study the effect of these on cambial activity and the differentiation of xylem and phloem. All three variables had highly significant effects on the total width of tissue produced during the experiment. However, the effects of GA were small except under conditions of high water potential and in the presence of IAA. Then the highest concentrations of GA (100 mg l^{-1}) were associated with the greatest number of cells. On the other hand, the highest concentrations of IAA (also 100 mg l^{-1}) had a depressing effect on division. A reduction in water potential at first led to a marked decrease in tissue width, and this effect occurred with practically all concentrations of the growth substances. Further reductions, however, had little effect except at high concentrations of GA with IAA present. Since wide ranges of hydrostatic pressure have been demonstrated in intact trees, this pronounced depressing effect of even slight reduction in water potential may be of considerable importance.

Environmental factors can influence the hormonal control of cambial activity. The effect of gravity will be discussed in the next chapter, but the effects of several environmental factors on cambial activity in roots have been investigated by Fayle (1968). Exposure to light was found to be by far the most influential factor, leading to a marked increase in cambial activity on the lower, less exposed, side.

This insight into the hormonal relationships of the activity of the cambium and the differentiation of its derivatives which has just been described cannot be regarded as going beneath the surface of these problems. Although encouraging beginnings have been made, much progress is needed before the intricate factors of the internal environment which control the morphogenesis of such precise and specific anatomical patterns as occur in secondary tissues are understood. Within the cambium itself a balance between radial and circumferential growth is maintained and the ratio of ray initials to the total area of cambium is kept at levels characteristic of the species. The control of the differentiation of the derivatives as they form the complex tissues of phloem and xylem is outside the scope of this book, but it cannot be doubted that, for example, such woods as *Hoheria*, in which a single growth ring may be made up of twenty or more zones of fibres alternating with as many zones of parenchyma and vessels, offer challenging material for experimental study. Some encouraging results of the application of growth substances to differentiating xylem elements are included in the next chapter, in which the influence of gravity on the cambium and its derivatives is considered.

Important advances towards understanding the fundamental problems of differentiating vascular tissues, and the cambium, were reported by Wetmore and Rier (1963) and further discussed by Wetmore *et al.* (1964). Their results support the view that auxin and sugar are critical variables in the induction and differentiation of vascular tissues. If sucrose and auxin at appropriate concentrations are applied in agar to the surface of a callus culture, together they induce and influence the development of vascular elements. Without these substances controls remain healthy but virtually without vascular elements. If the level of applied auxin is held constant varying the concentration of sucrose will alter the balance of xylem and phloem production. At lower concentrations (1·5–3%) xylem results, at 4·5–5% phloem is produced, and at concentrations between these values both types of vascular tissue tend to appear, often with a cambial layer between. The vascular elements occur as a ring of nodules 1·5–1·7 mm below the applied agar. The small sectors of cambium within the nodules may eventually join by the formation of interfascicular arcs of cambium until a partial or even a complete circle of cambium may develop. Increasing the concentration of the applied auxin has the effect of increasing the diameter of the ring formed by the induced nodules of vascular tissue. It is suggested that, as in these callus tissue-cultures, sugar may well prove to be an important variable concerned in the induction and subsequent differentiation of the vascular tissues of intact plants. It is also important to recognize that the isodiametric cells of homogeneous callus tissue are as suitable a substrate for the induction of vascular tissues as are the elongated cells of procambium, which generally contrast with the other cells in the apices of angiosperms. The induction of phloem and xylem occur if and when the necessary biochemical inducing agents are present in appropriate concentrations regardless of cell shape. That this is true also of intact plants is evident, for example, in short shoots, where, in the absence of the usual elongation of cells, the xylem and phloem cells remain almost isodiametric.

Some interesting characteristics of the induction of cambium in roots are brought out by the work of Torrey and Loomis. Excised roots in sterile culture usually do not form a cambium or secondary vascular tissue. By using appropriate metabolite and hormonal mixtures which were introduced by way of the medium into the cut basal end of the root, Torrey (1963) induced a cambium in excised roots of the pea, and Loomis and Torrey (1964) applied this technique to the radish. More recently, Torrey and Loomis (1967a) have reported the effects of applying various concentrations of auxin and cytokinins to excised *Raphanus* roots. These experiments were undertaken to determine the optimum conditions for cambial initiation and for extended cambial activity. They found that an auxin (e.g. IAA at 10^{-6} M), a cytokinin (e.g. 6-benzylaminopurine at 5×10^{-6} M), a cyclitol (such as *myo*-inositol at 5×10^{-4} M) and sucrose at 8% were all required for maximum response. In the absence of auxin and cytokinin no cambium was formed. Cyclitol was not an absolute requirement, but with it the response was strikingly enhanced. Some alternative auxins, cytokinins, and cyclitols were found to be equally effective, whereas others were less effective or totally inactive. The cell divisions induced in this way within the excised root occur in an organized tissue system at a place and in a sequence not unlike those which initiate the cambium in an intact plant. The authors, therefore, suggest that the shoot supplied the same kind of hormones in similar balance to the root. They consider that the auxin functions through activating divisions in the procambial cells. The events which follow, leading to the differentiation of xylem and phloem elements, are no doubt explicable in terms of the influences of preexisting root structure rather than of the hormones supplied. Torrey and Loomis (1967b) have compared the details of ontogeny of the vascular cambium in radish roots cultured in optimal conditions with that of intact seedlings grown in 16-hr days and found them to be similar.

The undoubted importance of the pattern of hormonal levels in controlling cambial activity and the development of secondary tissues should not lead to other factors being overlooked. Brown (1964) has shown the importance of pressure on the differentiation of cambial derivatives. Explants of strips of the inner phloem and cambium of *Populus deltoides* Marsh. developed callus by the proliferation of all the surface cells. No regular cambial-like divisions persisted and no organized differentiation into xylem and phloem occurred.

An ingenious technique allowed pressures of 0·05–1·0 atm. to be applied to the surfaces of similar explants while in sterile culture. In these cultures the cambium remained intact, and its derivatives, at least after an initial check, differentiated into normal elongate xylem elements.

Studies of cambial regeneration are also of value in understanding factors which control the site of cambial activity. If a cambial ring is partially destroyed by a deep wound the ring will be restored by the regeneration of cambial activity in the wound callus. Wilson and Wilson (1961) have discussed the nature of these regeneration cambia and put forward an hypothesis on the factors which must govern their position. It seems likely that the initiation of the cambium is determined by gradients which may establish under the influence either of conditions external to the wound or of pre-existing vascular tissues, or of both. The orientation of the cambium with respect to xylem and phloem production is determined by the direction of this gradient.

Another interesting type of experiment is that performed by Tupper-Carey (1930). She ring-barked trees of *Laburnum* and *Acer*, but left bridges of tissue with horizontal portions linking the bark above and below the cut. At first the lack of pressure within these bridges resulted in the formation of callus-like tissue and the cambial initials by repeated divisions came to resemble ray cells. At a later stage some of this mass of isodiametric cells became elongated horizontally, i.e. in the direction of the bridge tissue. In this way tracheary elements were formed with their axis parallel to the flow of any materials being transported across the bridges.

REFERENCES

AVERY, G. S., BURKHOLDER, P. R. and CREIGHTON, H. B. (1937). Production and distribution of growth hormone in shoots of *Aesculus* and *Malus* and its probable role in the stimulation of cambial activity. *Am. J. Botany* **24**, 51–8.

BROWN, A. B. (1936). Cambial activity in poplar with particular reference to polarity phenomena. *Can. J. Res. C.* **14**, 74–88.

BROWN, C. L. (1964). The influence of external pressure on the differentiation of cells and tissues cultured in vitro. In ZIMMERMANN, M. H., *The Formation of Wood in Forest Trees*, Academic Press, New York.

DE MAGGIO, A. E. (1966). Phloem differentiation: induced stimulation by gibberellic acid. *Science, N.Y.* **152**, 370–2.

DIGBY, J. and WAREING, P. F. (1966a). The effect of applied growth hormones on cambial division and the differentiation of the cambial derivatives. *Ann. Botany, N.S.* **30**, 539–48.

— and — (1966b). The relationship between endogenous hormone levels in the plant and seasonal aspects of cambial activity. *Ann. Botany, N.S.* **30**, 607–22.

DOLEY, D. and LEYTON, L. (1968). Effects of growth regulating substances and water potential on the development of secondary xylem in *Fraxinus*. *New Phytol.* **67**, 579–94.

FAYLE, D. C. F. (1968). Radial growth in tree roots. *Fac. For. Univ. Toronto*, Tech. Rept. No. 19, 1–183.

JOST, L. (1893). Über Begiehungen zwischen der Blattentwicklung und der Gefässbildung in der Pflanze. *Botan. Z.* **51**, 89–138.

LOOMIS, R. and TORREY, J. G. (1964). Chemical control of vascular cambium initiation in isolated radish roots. *Proc. Natl Acad. Sci. (U.S.)* **52**, 3–11.

MOREY, P. R. and CRONSHAW, J. (1968). Developmental changes in the secondary xylem of *Acer rubrum* induced by gibberellic acid, various auxins and 2,3,5-triiodobenzoic acid. *Protoplasma* **65**, 315–26.

OKUNDA, M. (1959). Response of *Pharbitis nil* Chois to gibberellin with special reference to anatomical features. *Botan. Mag., Tokyo* **72**, 443–4.

REINDERS-GOUWENTAK, C. A. (1949). Cambiumwerkzaamheid en groeistof. *Vakbb. Biol.* **29**, 9–17.

— (1965). Physiology of the cambium and other secondary meristems of the shoot. In RUHLAND, W. (ed.), *Encyclopedia of Plant Physiology* XV (1), pp. 1077–105, Springer, Berlin.

SHEPHERD, K. R. (1964). Some observations on the effect of drought on the growth of *Pinus radiata*. *Austr. For.* **28**, 7–22.

— and ROWAN, K. S. (1966). Indoleacetic acid in cambial tissue of Radiata pine. *Australian J. Biol. Sci.* **20**, 637–46.

SNOW, R. (1933). The nature of the cambial stimulus. *New Phytol.* **32**, 288–96.

— (1935). Activation of cambial growth by pure hormones. *New Phytol.* **34**, 347–60.

SÖDING, H. (1940). Weitere Untersuchungen über die Wuchsstoffregulation der Kambiumätigkeit. *Z. Bot.* **36**, 113–41.

TORREY, J. G. (1963). Cellular patterns in developing roots. *Symp. Soc. Exp. Biol.* **17**, 285–314.

— and LOOMIS, R. S. (1967a). Auxin–cytokinin control of secondary vascular tissue formation in isolated roots of *Raphanus*. *Am. J. Botany* **54**, 1098–106.

— and — (1967b). Ontogenetic studies of vascular cambium formation in excised roots of *Raphanus sativus* L. *Phytomorphol.* **17**, 401–9.

TUPPER-CAREY, R. M. (1930). Observations on the anatomical changes in tissue bridges across rings through the phloem of trees. *Proc. Leeds Phil. Lit. Soc. (Sci. Sect.)* **2**, 86–94.

WAREING, P. F. (1951). Growth studies in woody species. IV. The initiation of cambial activity in ring-porous species. *Physiol. Plantarum* **4**, 546–62.

—, HANNEY, C. E. A. and DIGBY, J. (1964). The role of endogenous hormones in cambial activity and xylem differentiation. In ZIMMERMANN, M. H. (ed.), *The formation of Wood in Forest Trees*, pp. 323–44, Academic Press, New York.

— and ROBERTS, L. W. (1956). Photoperiodic control of cambial activity in *Robinia pseudoacacia* L. *New Phytol.* **55**, 356–66.

WETMORE, R. H., DE MAGGIO, A. E. and RIER, J. P. (1964). Contemporary outlook on the differentiation of vascular tissues. *Phytomorphol.* **14**, 203–17.

WETMORE, R. H. and RIER, J. P. (1963). Experimental induction of vascular tissues in callus of angiosperms. *Am. J. Botany* **50,** 418–30.

WHITMORE, F. W. (1968). Auxin and cell wall formation in the cambial derivatives of Pine and Cottonwood. *Forest Sci.* **14,** 197–205.

— and FAHNER, R. (1966). Development of the xylem ring in stems of young red pine trees. *Forest Sci.* **12,** 198–210.

WILSON, J. W. and WILSON, P. M. (1961). The position of regenerating cambia – a new hypothesis. *New Phytol.* **60,** 63–73.

11
Reaction of the cambium to gravity and to the displacement of branches

The asymmetrical growth of stems has already been mentioned in connection with anomalous forms of secondary thickening (Chapter 6). Eccentric growth also occurs frequently in branches or trunks which are inclined to the vertical. In addition to this asymmetry, inclined stems often possess tissues anatomically different from those of vertical shoots; these are referred to as reaction tissues.

In dicotyledons well-defined reaction wood is characterized by fibres in which the typical series of layers of the secondary wall is modified by the presence of a gelatinous layer. This layer is unlignified, or only slightly lignified, and the cellulose microfibrils are oriented axially. The average fibre length is usually shorter than in equivalent normal wood, and fewer vessels may be present. Many woody dicotyledons are known in which gelatinous fibres do not occur (Onaka, 1949; Patel, 1964; Scurfield, 1964), and some trees develop reaction xylem with distinctive features (Onaka, 1949). The reaction wood of *Entelea arborescens* R.Br., for example, lacks gelatinous fibres but differs from the normal secondary xylem of that species in the absence of specialized parenchyma. In *Aristotelia*, also, gelatinous fibres are absent, but the reaction xylem differs in several respects from normal wood, especially in the much higher rays (unpublished data). Gelatinous fibres do not occur in the reaction wood of gymnosperms, though they have been reported from the phloem (Liese and Höster, 1966). Gymnospermous reaction wood is characterized by tracheids which are rounded in transverse section, so that conspicuous air spaces appear between them. A full account of the anatomical features of reaction wood is inappropriate here, but will be found in Wardrop (1964, 1965), White (1965), Westing (1965a), and Côté and Day (1965). The asymmetry of the growth rings is characteristically different in dicotyledons and gymnosperms. In the former growth is accentuated on the upper side of a leaning axis, whereas in gymnosperms it is on the lower side. The eccentricity in both these plant groups is normally due to an acceleration of cambial activity on one side, with a corresponding decrease in radial growth on the opposite side. The reaction wood of *Gnetum* is exceptional, since it is similar to that of dicotyledons (Westing, 1965a).

It is usual for the anatomically distinctive wood to occur on the same side as the exaggerated growth (Robards, 1965). That is to say, the upper side of a dicotyledonous branch has thicker rings composed of reaction-type wood, whereas the underside of a gymnospermous branch is enlarged and is composed of modified tracheids. The reaction wood of dicotyledons is therefore often referred to as tension wood and that of gymnosperms as compression wood. However, accentuated growth and anatomical modification do not always occur together in this way. White (1962) reported that much more growth occurred on the lower side of a branch of *Sassafras officinale* Nees and Eburn., in which the gelatinous fibres were on the upper side, and this is a constant feature of lateral branches of the vessel-less dicotyledon *Pseudowintera colorata* (Raoul) Dandy (unpublished data). Dadswell and Wardrop (1949) report a similar situation near the base of lateral branches of silver wattle, and Robards (1965) found that tension wood formed both above and below in most growth rings from the base of a branch of *Salix fragilis* L. Irregularities of this kind may be characteristic of branch bases, because in *Hoheria* growth may be accentuated on the lower side near the base of inclined branches, although for the rest of the branch the typical dicotyledonous activity occurs (unpublished data). Onaka (1949) lists several dicotyledons in which eccentric growth may be on the lower side as in gymnosperms, and Patel (1964) discusses the variable position of the pith in horizontal branches. Further, reaction anatomy may occur with no asymmetry of the growth rings, and many woody plants are not known to produce reaction wood at all. The unusual reaction wood of *Entelea* and *Aristotelia* referred to above occurs on the underside of inclined branches, that is on the side with least radial growth (Fig. 11.1). In arborescent monocotyledons displaced branches will bend upwards by a growth curvature in the primary tissues of the apex (Scurfield, 1964). These displaced branches show an accentuated lignification of the bundle sheaths. There is an increase in the number of vascular elements in those displaced bundles, and their lignification is accentuated, but they appear to be incapable of corrective movements in the zone of secondary growth.

Clearly, then, there are at least two separate reactions involved in the development of reaction wood. One affects the rate of tangential division in the cambium, the other the nature of the secondary wall of fibres and tracheids, besides several other modifications in the developing xylem and phloem.

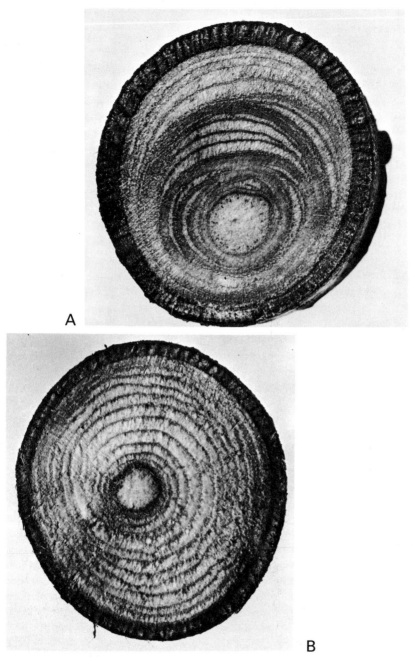

Fig. 11.1 Entelea arborescens. T.S. of branches. A, through horizontal stem. On the lower side parenchyma formation has been suppressed. B, through approximately vertical stem. The secondary xylem consists of alternating bands of fibres and parenchyma.

That the formation of reaction wood is usually a response to the stimulus of gravity can be concluded from the results of several experiments. Reaction wood may form in vertical stems under natural conditions, and may be related to slight deflections, perhaps due to uneven growth, to which the stems have been exposed. Its distribution around the stem at different heights is often spiral or irregular. Robards (1966) has shown how very slight deflections of the stem of *Salix fragilis* result in the differentiation of gelatinous fibres. The amount formed in inclined stems of equal vigour is dependent on the angle of lean, being at a maximum when the stem is inclined at about 120° from the upright (Robards, 1966). Seedlings of *Pinus* and *Larix* which were tilted from the vertical and then restored to an upright position on several occasions were found to have developed a corresponding succession of arcs of reaction xylem (Kennedy and Farrar, 1965). In gymnospermous stems which are bent into a complete loop the 'compression' wood forms on the lower side of the stem at both the top and bottom of the loop. This is in spite of the fact that at the top of the loop the wood is under compression, while at the bottom it is under tension. Dicotyledons show the same effect, but with the positions of the reaction wood reversed. Stems placed horizontally on a rotating clino- stat form no reaction wood even if bent, provided that the speed of rotation is relatively high (e.g. one revolution in less than about $\frac{1}{2}$ hr). If one rotation takes about 1 hr or longer traces of reaction wood occur all round the stem. A very conclusive experiment was performed by Burns (1942). He grew pine trees horizontally and bent their stems sideways. Thus one side of the stem was under tension and the other under compression. Nevertheless, the re- action wood formed on the underside of the stem, including the bent portion, and therefore could not be a response to stresses. Jaccard (1939) showed that centrifugal force could replace gravity in the induction of reaction wood.

The examples given above refer to stems which are normally upright and which react to a displacement. Similar responses occur in naturally drooping terminal shoots (Mergen, 1958) or in seedling trees with initially semi- prostrate main shoots (Little and Mergen, 1966). Lateral branches which are inclined at an angle normally develop similar reaction wood, and the amount of it will be increased if the branch is depressed below its normal angle of rest. However, if such a lateral branch be displaced upwards the reaction wood will form on the opposite side (Hartmann, 1949). All these effects can be unified by the hypothesis that reaction wood forms in the position which would tend to restore the axis to its normal position. Wardrop (1964) sug- gests that 'the degree of reaction wood formation and its distribution in the branch are related to the sign of the geotropic movement on the part of the branch in which it occurs'. Westing (1965a) postulates a type of statolith to account for these responses. The maintenance of a characteristic angle of branching is an expression of the correlation of growth in different parts of the tree. If the symmetry of the tree is destroyed by loss of a member other members may respond by the development of reaction wood accompanied by movement. The most familiar example is the replacement of a leader by lateral branches. Westing (1965b) has shown that the lateral spacing of members may be controlled in the same way. When one branch was removed from a whorl of *Pinus radiata* the adjacent branches moved together slightly, closing the gap, and appropriate development of reaction wood was observed. The stimulus of gravity can have no part in this response.

Some evidence suggests that where gravity acts as the stimulus it may induce these asymmetrical responses of the cambium through its effect on the distribution of hormones in inclined stems. Wareing *et al.* (1964), in experiments with *Populus robusta*, made slits in the apical ends of pieces cut from erect dormant shoots and inserted cover slips to separate the two halves of the stems. IAA and GA alone and together were applied to either the upper or lower halves of these twigs, which were then kept horizontal. After three weeks the twigs were sectioned at 1, 3, and 5 cm from the treated ends, and the increments of xylem on the upper and lower sides were measured. The

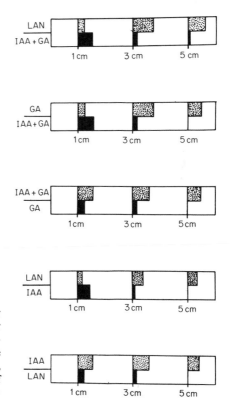

Fig. 11.2 Effect of applying IAA and/or GA to the apical ends of horizontal poplar shoots. The stippled and black areas indicate the mean width of new xylem tissue on the upper and lower sides, respectively, at various distances from the point of application. (From Wareing *et al.*, 1964.)

results are recorded in Fig. 11.2. Within 1 cm of the treated end more cambial activity occurred in the half which had received hormone. Farther from the end, however, more activity occurred in the upper half, no matter where the hormone had been applied. The absence of cambial activity on the lower side of these horizontal shoots at some distance from the apical end might be due either to a lack of auxin or to inhibitory levels of auxin. To resolve this doubt, additional applications of auxin were made to cuts along the lower surface of the twigs. Since cambial activity occurred near these applications, it was concluded that the original lack of activity was due to a deficiency of IAA; a similar experiment using the gymnosperm *Pseudostuga taxifolia* (Poir.) Britton resulted in greater cambial activity on the lower side in all cases, no matter whether hormones were applied to the upper or lower half. These results suggest that besides the polar transport of auxin there is present a mechanism for the redistribution of auxin in horizontal shoots. Since auxin is known to stimulate cambial activity, it would seem that in dicotyledonous trees it is redistributed in such a way that there are greater concentrations on the upper side, whereas in conifers the reverse is true. However, later work from the same laboratory (Leach and Wareing, 1967) established that the concentration of auxin in horizontal shoots is greater on the lower than the upper side. The lesser activity of the cambium on the lower side of dicotyledonous branches is explained by the presence of inhibitors which also were found to be concentrated on the under side of horizontal stems.

While these results form some basis for explaining one effect of gravity, i.e. differential activity, they do not apply to the anatomical differentiation so characteristic of reaction wood. It has long been known that compression wood can be induced even in upright shoots of conifers by the application of auxin (Wershing and Bailey, 1942). But this has not usually been found possible with dicotyledons (Gouwentak, 1936). Indeed, Nečesaný (1958) found that applications of IAA to the upper side of bent poplar stems inhibited the formation of reaction wood. He concluded that a high concentration of IAA promotes the anatomical characteristics of compression wood in gymnosperms, but that a low concentration is required for the features of tension wood to be developed in dicotyledons. This is confirmed by the induction of tension wood by the anti-auxin 2,3,5-tri-iodobenzoic acid (Cronshaw and Morey, 1965; Kennedy and Farrar, 1965).

Further progress in understanding the factors controlling the two aspects of reaction wood, namely eccentric growth and modified differentiation, has been made by the investigations of Morey and Cronshaw (1968, a, b, c). Tension wood was induced in internodes of upright seedlings of *Acer rubrum* by applications of both 2,3,5-tri-iodobenzoic acid and 2,4-dinitrophenol. Both of these substances are known to interfere with the polar transport of auxin. Simultaneous applications of auxins counteract the induction of tension wood. It is concluded that tension wood forms in conditions of auxin deficiency, in these experiments caused by local blocking of polar transport. Enhanced cambial activity and tension wood formation are normally associated in the eccentric development of tension wood in horizontal branches. It is difficult to reconcile these two features of gravitational response, since one is related to the presence of auxin and the other to its deficiency. It is known that gibberellic acid stimulates cambial activity as well as auxin (see last chapter). But unlike auxin, gibberellic acid has no direct morphogenetic effect on the development of secondary xylem. Morey and Cronshaw (1968b) confirm both the stimulating effect of GA on cambial activity, provided the endogenous rate is low, and also its lack of morphogenetic effect on developing xylem. They point out that the conditions for the formation of both aspects of reaction to gravity would be present if there were low levels of auxin and high levels of gibberellin. Tension-wood anatomy could be expected due to the low concentration of auxin, while radial growth would be accelerated by the gibberellic acid. Simultaneous application of GA with an auxin antagonist resulted in larger amounts of tension wood, but it is true that the stimulation of the cambium was only slight in the conditions of their experiments. Nevertheless, their work represents a great advance in the understanding of reaction wood.

During their experiments with *Acer rubrum* seedlings Morey and Cronshaw frequently recorded the induction of tension wood by various auxins, especially at low concentrations. They explain this apparent inconsistency (Morey and Cronshaw, 1968c) by suggesting that the applied auxin stimulated cambial activity, and this reduced the amount of auxin present on the xylem side of the cambium by the growth of the increased derivatives. The pathway of differentiation of these derivatives will thus be switched from that of libriform fibres to that of tension-wood fibres.

Kennedy and Farrar (1965) were able to demonstrate that the anatomical response to gravity of developing tracheids in compression wood was dependent on the stage in their development. The least differentiated cambial cells were stimulated to divide, resulting in the typical eccentric growth of the inclined stem. Cells affected in the latest phase of differentiation produced tracheids which differed from the normal only in the intensity of their lignification. Cells in an earlier phase responded by having thicker walls as well as heavier lignification. The typical rounded form of the compression fibre resulted only if the cell was affected during its early period of enlargement. Thus all the features of a typical compression fibre were developed only if the differentiating cells were on the under side of the stem for the entire period of their development.

Westing (1965a) concedes that asymmetry of auxin concentration occurs in inclined branches, but doubts whether it is sufficiently marked to account for the differences in cambial activity and differentiation required for reaction wood production. Very slight upward or downward movements of a stem will induce response, whereas the effect on auxin concentration is unlikely to be significant. Westing inclines to the hypothesis that the stimulation induces differential sensitivity to auxin. It is known that a basipetal factor, presumably auxin, is required for the development of reaction wood (Wardrop, 1964). This local sensitization to auxin might be brought about by the asymmetrical production of an immobile substance. That reaction wood can be induced by the presence of such a substance is shown by the development of compression-type xylem in the galls of *Abies balsamea* (L.) Mill., which result from injections made by the woolly aphid *Adelges piceae* (Balch *et al.*, 1964). However, Smith (1968) has pointed out several respects in which this gall tissue differs from normal compression wood.

The perception of the stimulus of gravity is almost certainly close to the location of the response, because this is independent of the orientation of adjacent portions of the axis, and particularly of the direction of growth of the apex. That perception occurs in the cambium is not improbable, since phloem as well as xylem is affected.

Some experiments reported by Casperson (1960) demonstrate the importance of the cambium or its immediate derivatives in the perception of the stimulus, and possibly also as the tissue responsible for the resultant movements of readjustment. Seedlings of the Horse-chestnut (*Aesculus*) showed movement responses within one day of exposure to the stimulus of gravity. In these plants the first reaction wood fibres did not appear until cells present in the cambial zone at the time of stimulation had reached maturity: that is to say, after an interval of about ten days. Casperson concluded that gelatinous fibres resulted from a stimulus affecting cambial cells in such a way that their subsequent differentiation was altered, and further that the reaction movements of plants were not due to the presence of reaction wood. These results lend support to the view of Frey-Wyssling (1952) that growth forces within the cambium can cause reaction movements of woody stems. But it is possible that the mechanism involved in the bending of these seedlings, before the differentiation of reaction wood, is different from that functioning in more mature woody stems. Twigs of woody plants while still in the primary phase of growth, like the stems of herbs, are known to develop growth curvatures in response to gravity by mechanisms which do not involve gelatinous fibres. It is possible that these also operate in these seedlings before secondary wood has matured. While opinion remains divided on the mechanism of these movements, most authorities regard the relatively great stresses present in reaction wood as responsible, and Hejnowicz (1967) presents evidence that these stresses are related to imbibition of water by the cell walls of the reaction fibres. It should be borne in mind, however, that orientation movements by hardwoods which lack gelatinous fibres suggest that the mechanisms involved may vary (Scurfield, 1964). While the forces developed within reaction wood are not cambial, and are therefore not relevant to this book, this account of reaction tissues may end by reference to the reviews by Westing (1965a, 1968), in which the mechanism of righting is discussed. The force required to prevent the righting of a displaced sapling increases with time, and consequently with the increasing development of reaction wood. Cronshaw and Morey (1968) found a close relationship between geotropic reorientation and the amount of tension wood formed in *Acer rubrum* seedlings treated with various concentrations of auxins, and support the view that tension wood fibres in the wall-thickening stage of their differentiation are an active component of the mechanism of righting. While the effect of differential increases in the osmotic forces in the cambium is not ruled out, it is regarded at best as only a minor factor.

It may be concluded, therefore, that the cambium receives the stimulus of gravity or displacement, and itself reacts by the differential radial activity which leads to eccentric growth. The anatomical features of reaction tissues, however, are not determined by changes induced in the cambium, but are the result of the stimulus acting on differentiating vascular elements. The result is dependent on the stage of development of the element at the time it is affected by the stimulus.

REFERENCES

BALCH, R. E., CLARK, J. and BONGA, J. M. (1964). Hormonal action in production of tumerous and compression wood by an aphid. *Nature, Lond.* **202**, 721–2.

BURNS, G. P. (1942). Excentric growth and the formation of redwood in the main stem of conifers. *Vermont Agric. Expt. Sta., Bull.* **219**, 1–16.

CASPERSON, G. (1960). Uber die Bildung von Zellwänden bei Laubhölzern, I, Mitteilung. Feststellung der Kambiumaktivität durch Erzuegen von Reaktionshölz. *Ber. Deutsch Bot. Ges.* **73**, 349–57.

CÔTÉ, W. A. and DAY, A. C. (1965). Anatomy and ultrastructure of reaction wood. In CÔTÉ, W. A., *Cellular Ultrastructure of Woody Plants*, Syracuse University Press, Syracuse.

CRONSHAW, J. and MOREY, P. R. (1965). Induction of tension wood by 2,3,5-tri-iodobenzoic acid. *Nature, Lond.* **205**, 816–18.

— and — (1968). The effect of plant growth substances on the development of tension wood in horizontally inclined stems of *Acer rubrum* seedlings. *Protoplasma* **65**, 379–91.

DADSWELL, H. E. and WARDROP, A. B. (1949). What is reaction wood? *Australian For.* **13**, 22–33.

FREY-WYSSLING, A. (1952). Wachstumleistungen der pflanzenlichen Zytoplasmas. *Ber. Schweiz Bot. Ges.* **62**, 583–91.

GOUWENTAK, C. A. (1936). Kambiumtätigkeit und Wachstoff. *Meded. Landbouwhoogesch.* (Wageningen) **39**, 3–23.

HARTMANN, F. (1942). *Das statische Wuchsgesetz bei Nadel- und Laubbäumen. Neue Erkenntnis über Ursache, Gesetzmässigkeit und Sinn des Reaktionsholzes,* Springer, Vienna.

HEJNOWICZ, Z. (1967). Some observations on the mechanism of orientation movements of woody stems. *Am. J. Botany* **54**, 684–9.

JACCARD, P. (1939). Tropisme et bois de reaction provoque par la force centrifuge. *Bull. Soc. Bot. Suisse* **49**, 135–47.

KENNEDY, R. W. and FARRAR, J. L. (1965). Tracheid development in tilted seedlings. In CÔTÉ, W. A., *Cellular Ultrastructure of Woody Plants,* Syracuse University Press, Syracuse.

— and — (1965). Induction of tension wood with the anti-auxin 2,3,5-tri-iodobenzoic acid. *Nature, Lond.* **208**, 406–7.

LEACH, R. W. A. and WAREING, P. F. (1967). Distribution of auxin in horizontal woody stems in relation to gravimorphism. *Nature, Lond.* **214**, 1025–7.

LIESE, W. and HÖSTER, H. R. (1966). Gelatinöse bastfasern im Phloem einiger Gymnospermen. *Planta* **69**, 338–46.

LITTLE, S. and MERGEN, F. (1966). External and internal changes associated with basal-crook formation in Pitch and Shortleaf Pines. *Forest Sci.* **12**, 268–75.

MERGEN, F. (1958). Distribution of reaction wood in eastern hemlock as a function of its terminal growth. *Forest Sci.* **4**, 98–109.

MOREY, P. R. and CRONSHAW, J. (1968a). Developmental changes in the secondary xylem of *Acer rubrum* induced by various auxins and 2,3,5-tri-iodobenzoic acid. *Protoplasma* **65**, 287–312.

— and — (1968b). Developmental changes in the secondary xylem of *Acer rubrum* induced by gibberellic acid, various auxins and 2,3,5-tri-iodobenzoic acid. *Protoplasma* **65**, 315–26.

— and — (1968c). Induction of tension wood by 2,4-dinitrophenol and auxins. *Protoplasma* **65**, 393–405.

NEČESANÝ, V. (1958). Effect of β-indoleacetic acid in the formation of reaction wood. *Phyton* **11**, 117–27.

ONAKA, F. (1949). Studies on compression and tension wood. *Bull. Wood. Res. Inst. Kyoto Univ.* **1**, 1–83.

PATEL, R. N. (1964). On the occurrence of gelatinous fibres with special reference to root wood. *Inst. Wood. Sci.* **12**, 67–80.

ROBARDS, A. W. (1965). Tension wood and eccentric growth in Crack Willow (*Salix fragilis* L.). *Ann. Bot.* **29**, 419–31.

— (1966). The application of the modified sine rule to tension wood production and eccentric growth in the stem of Crack Willow (*Salix fragilis* L.). *Ann. Bot.* **30**, 513–23.

SCURFIELD, C. (1964). The nature of reaction wood. IX. Anomalous case of reaction anatomy. *Australian J. Botany* **12**, 173–84.

SMITH, F. H. (1968). Effects of balsam woolly aphis (*Adelges piceae*) infestation on cambial activity in *Abies grandis*. *Am. J. Botany* **54**, 1215–23.

WARDROP, A. B. (1964). The reaction anatomy of arborescent angiosperms. In ZIMMERMANN, A. H., *The Formation of Wood in Forest Trees*, Academic Press, New York.

— (1965). The formation and function of reaction wood. In CÔTÉ, W. A., *Cellular Ultrastructure of Woody Plants*, Syracuse University Press, New York.

WAREING, P. F., HANNEY, C. E. A. and DIGBY, J. (1964). The role of endogenous hormones in cambial activity and xylem differentiation. In ZIMMERMANN, M. H., *The Formation of Wood in Forest Trees*, Academic Press, New York.

WERSHING, H. F. and BAILEY, I. W. (1942). Seedlings as experimental material in the study of 'redwood' in conifers. *J. Forest.* **40**, 411–14.

WESTING, A. H. (1965a). Formation and function of compression wood in gymnosperms. *Botan. Rev.* **31**, 381–480.

— (1965b). Compression wood in the regulation of branch angle in gymnosperms. *Bull. Torrey Bot. Club* **92**, 62–6.

— (1968). Formation and function of compression wood in gymnosperms. II. *Botan. Rev.* **34**, 51–78.

WHITE, D. J. B. (1962). Tension wood in a branch of Sassafras. *J. Inst. Wood Sci.* **10**, 74–80.

— (1965). The anatomy of reaction tissues of plants. In CARTHY, J. D. and DUDDINGTON, C. L., *Viewpoints in Biology, IV*, Butterworth, London.

Index

(Bold face type indicates figures)